Discovering Hedgerows

Discovering Hedgerows

David Streeter
Rosamond Richardson

The British Broadcasting Corporation

This book accompanies the BBC Television Series
Discovering Hedgerows, first broadcast on BBC2 from April 1982

Series presented by David Streeter and Rosamond Richardson
and produced by Erica Griffiths

Published to accompany a series of programmes
prepared in consultation with the
BBC Continuing Education Advisory Council

This book is set in 11 on 12 point Monophoto Apollo
Printed in England by Jolly & Barber Limited, Rugby
and bound by R. J. Acford Chichester Sussex

© The Authors
First Published 1982
Published by the British Broadcasting Corporation
35 Marylebone High Street, London W1M 4AA

Hardback ISBN 0 563 16452 2
Paperback ISBN 0 563 16528 6

Contents

Acknowledgements

The authors and the production team would like to express their gratitude to Johnathan Nott, Peter Fry, Christopher Morgan and Jim Foster for their kindness and tolerance in allowing access to their land throughout the seasons.

The authors would like to thank Valerie Green and Penny Streeter for typing the manuscript.

They would also like to say a special thankyou to Erica Griffiths, the producer of the BBC series which this book accompanies, for her constant encouragement, patience, tact and time and for her efforts in helping to edit together their separate contributions.

Illustrations
Cover and pages 17, 35, 53, 71, 89, 107 and 125 are by Julie Banyard of The Covent Garden Studio, London.
Pages 24, 25, 26, 27, 30, 33, 44, 49, 64, 65, 69, 75, 76, 80, 81, 82, 84, 85, 92, 93, 96, 98, 99, 110, 113, 114, 116, 128, 129 and 150 are by Joanna Langhorne MSIAD.
Pages 38, 78, 79, 136 and 137 are by Maggie Raynor.
Pages 16, 34, 39, 42, 43, 52, 70, 88, 106 and 124 are by Tony Spaul.

Introduction

Hedgerows are rooted deeply in our national consciousness as indispensable elements of our ideal of the English countryside. They stitch together the fabric of our landscape to create a scene unmatched except where newly created by seventeenth-century colonists in a few parts of the old Empire such as New England, Tasmania and parts of New Zealand. Many are more ancient by centuries than our castles, cathedrals and abbeys, and, in the words of W. G. Hoskins, 'they represent the physical evidence of decisions made long ago and fixed solidly on the ground'. By their character they proclaim the regional identity of the countryside as surely as a knight's pedigree was revealed by the charge on his shield.

But how much do we really know about our own favourite hedgerow? How old is it? Who created it? Why? What was it used for? How is it managed? Does it have an interesting natural history? Which of its herbs and berries can be used for cooking? What of its future? In this book we set out to answer some of these questions by looking at two hedges which are well known to us and by following their fortunes throughout the year from spring to autumn.

One is in the intensively cultivated arable farmland of north Essex and the other in the heavily wooded countryside of the Sussex High Weald. Our two hedges reflect these geographical differences, being creatures of their own landscapes. Although we believe that much of what we found will strike a familiar chord over a large area of lowland England, there inevitably remains a distinct flavour of the South-east. We have, therefore, spiced our pot with excursions to the South-west, the Welsh marches, the Cotswolds, and the North Yorkshire moors. Each month we concentrate on some aspect of the hedge's history and folklore, and provide ideas for recipes based on the natural hedgerow harvest. We also keep watch on the plant and animal life of the hedge as it responds to the changing seasons. Throughout, our aim has been to try and answer questions about the familiar sights of the ordinary hedgerow, from spiders' webs to cuckoo-spit. We have also tried to show how one can go about the fascinating business of discovering the history of our own particular hedges and the part that they have played in fashioning the local landscape.

What is a hedge? Surprisingly, that rather obvious question will produce quite different answers in different parts of the

Laying a Devon hedge on top of a turfed bank

country. In parts of Cornwall a 'hedge' is a low stone wall, some-times of tremendous antiquity, encircling a small, irregularly shaped field. In Devon a 'hedge' is a high, turfed bank surmounted by a line of living shrubs. In south-west Ireland the roadsides are lined by hedges of fuchsia; the Georgian enclosure landscape of the east Midlands is characterised by carefully-managed hedges of hawthorn; holly hedges are common in Staffordshire; while the open and exposed Breckland landscape of north-west Suffolk and south-west Norfolk is dominated by tall shelter belts of pine. In the Somerset levels, hedges of osiers stand up above the winter flood-waters to mark the boundaries of the submerged fields, while the Pennine hills of northern England are festooned with a jigsaw pattern of stone walls. The hop gardens of Kent and Worcestershire are protected by carefully tended, tall, dense shelter hedges, often more than 6 metres high, as also are the famous bulb-fields of the Isles of Scilly. However, for most people a hedge is a line of low shrubs up to about 2 metres wide and trimmed to between 1·2 metres and 1·8 metres high which forms the boundary to a garden or field, or borders a roadside. Unmanaged and left to their own devices, hedges become overgrown in a very few years. This can sometimes lead to a rather fine distinction between a broad hedge and a long, narrow wood. In parts of the Weald, for instance, the fields are typically bordered by narrow strips of woodland, locally called 'shaws', and some may consider it stretching the definition too far to regard these as true hedges. However, as we shall see, the question is rather academic, especially when viewed in the context of the hedge's history. These variations in the regional character of hedges are primarily reflections of differences in function and the availability of materials and plants with which to construct the hedge.

8

Above
Typical country lane and hedgerows in the Sussex Weald
Right
Dry stone walls replace traditional 'live' hedges as field boundaries in North Yorkshire

Historically, hedges served two main purposes: to delineate a political or territorial boundary and to provide a stock-proof barrier. Our first documentary references to hedges are to be found in Anglo-Saxon charters. In the absence of maps, boundaries to parishes, manors and property were described in great detail by reference to obvious landmarks, such as roads, streams and conspicuous trees. Dr Rackham has listed the trees mentioned in these early perambulations in order of the frequency of their use and, interestingly, the most commonly cited is 'thorn' – that is, hawthorn and sloe. The hawthorn is perhaps our commonest hedgerow shrub, although now it is less often met as a conspicuous, isolated tree. These early charters also contain references to hedges, and it is clear that the earliest hedges were by no means necessarily composed of living bushes. Indeed, the majority were probably 'dead hedges' of stakes, brushwood or stones, such as those that still surround the ancient Cornish fields. A dead hedge of stakes with thin, pliable branches woven between them is still built annually on the shore at Whitby as part of a tradition going back to the twelfth century. The story goes that in 1159 three knights and their hounds were pursuing a wounded boar in Eskdale when it went to ground in a hermitage which was at that time occupied by a monk of Whitby. The monk intervened to save the boar, and for his pains was promptly set upon by the hunters. The three escaped punishment at the hands of the Abbot of Whitby only because the dying monk forgave them, provided that they and the successors to their lands should each year, on the Eve of Ascension, gather stakes from the wood of Strayhead in Eskdale-side and carry them to the harbour at Whitby. 'Each of you shall set your Stakes at the Brim of the Water each Stake a Yard from another and so Yedder them, as with Yedders, and so stake on each side with your stout Stowers that they stand three tides without removing by the Force of the Water. . . . You shall do this Service in remembrance that you did cruelly slay me.' If this were not done their lands were to be forfeit to the Abbot. When we went to Whitby to see the building of this Penance or 'Penny' hedge, the stakes hadn't been gathered from the wood of Strayhead, but the hedge was built very much according to the monk's instructions and looked much as the early dead hedges must have done – although it has to be admitted that we did not stay long enough to confirm that it withstood the required three tides!

Pollard, Hooper and Moore in their indispensable book *Hedges* quote numerous contemporary documents from Anglo-Saxon custumals (manorial documents) to thirteenth-century records of abbey estates to conclude that dead hedges probably persisted as the main hedge type until well into the Middle Ages. The origins of live hedges as distinct from dead hedges is obscure. The late Ivan Margary found evidence of Roman field systems in the parishes of Ripe and Chalvington in Sussex, and similar patterns have been described from Brancaster in Norfolk. However, we have no means of knowing what kind of boundary surrounded the fields. As we have seen, clearance of waste and woodland for agriculture was well advanced by Saxon times, and it is quite possible that the shrubs and trees of the encircling forest edge were managed to produce a stock-proof barrier and hence a hedge. Alternatively, shrubs might have been gathered selectively from the wood and planted to form a hedge. What does seem clear is that our modern pattern of parish boundaries was already firmly established by the time of the Norman conquest, together with the open or common field system that formed the basis of the rural economy. Over large areas of lowland England a further period of active enclosure of the waste reached a peak in the thirteenth century. In some parts of the country, such as the Weald of Kent and Sussex, a pattern of small, irregularly shaped fields separated by broad hedges or thin strips of woodland are evidence today of this period of 'assarting', or the clearing of small areas of waste for agriculture. This time of agricultural activity was brought to an abrupt close by the Black Death in the 1340s, and for another two hundred and fifty years very little further change occurred in the English landscape. Thus, for those parts of lowland Britain not seriously affected by the Parliamentary enclosures, the present landscape was in most essentials already in existence by the middle of the fourteenth century.

Towards the end of the sixteenth century changes in agricultural practice began to favour smaller fields at the expense of the old open or common field system which had been in existence since before the Norman conquest over much of England, especially in the North-east and the eastern Midlands. Typically, the common field system consisted of the division of the land of each village into three large arable fields, one for fallow, one for winter corn and one for spring corn; a hay meadow and common grazing on rough land

The medieval open field system in the parish of Laxton, Nottinghamshire

or heath, moorland or sometimes woodland. There were numerous variations on this basic theme in different parts of the country, and some parishes might have possessed more than one manor each with its own or partially shared open field system. The parish of Laxton in Nottinghamshire still retains this ancient pattern almost intact. Some piecemeal enclosure had been achieved by the middle of the seventeenth century, but the remainder is still farmed on the original three-field system, until recently under the auspices of the Ministry of Agriculture, Fisheries and Food.

Some enclosure of the open fields had begun as far back as the mid-fifteenth century, and, as might be expected, enclosures frequently gave rise to disputes between the different parties involved, so that it became necessary to establish a legal basis for the action. A common ploy was to manufacture an entirely fictitious dispute and to take it to court. This would then result in the award of a Chancery Decree, so legally validating the enclosure. The Chancery Decree remained the most frequent means of effecting legal enclosures until the middle of the eighteenth century, when they were superseded by Parliamentary Acts. There were two main periods of Parliamentary enclosure. The first took place in the 1760s and 1770s when it has been estimated that about one and a half million acres were enclosed. The second and greatest surge of enclosure occurred between 1790 and 1815 and coincided with the dramatic increase in the price of farm produce as a result of the Napoleonic wars.

The reasons for enclosure were numerous and complex, and varied over the centuries in response to changes in the social and economic climate of the times. There were no doubt advantages in concentrating the land of a particular holding around the farm. The success of the common field system also depended heavily on the co-operative effort of the various tenants, so that advances in husbandry could be frustrated by a few conservative individuals. There were regional variations, too. Enclosure occurred early on the heavy clay soils of parts of the Midlands where the grazing system based on permanent pasture and long grass leys was not possible in the common field system. Many large landowners, forced into exile by Cromwell, returned at the Restoration imbued with new ideas in husbandry. Revolutionary developments in farming practice in the eighteenth century, especially the introduction of the Norfolk four-course rotation pioneered by Lord Townshend, further fuelled the movement for enclosure. The period of the Georgian enclosures coincided with the upsurge of the Industrial Revolution and the demand for cereal and potato crops to feed the new urban population. In the clothing trade cotton was replacing wool, reducing the large tracts of sheep ranch, and the Napoleonic wars raised the price of food generally.

The area of land affected by the Georgian enclosures was enormous, and involved a great swathe of the country stretching from the North-east, down through the Midlands to the southern counties from Dorset to West Sussex. The only areas of England hardly to be affected at all were the Welsh marches, the South-west and the south-eastern counties of Essex, Kent and East Sussex, including of course the areas in which our two hedges occur. Altogether some five thousand Acts were passed dealing with about seven million acres of land.

The Enclosure and Tithe Awards which resulted from the Acts frequently set out in great detail how the new proprietors were to hedge and ditch their new boundaries. For example, Pollard, Hooper and Moore quote: 'plots of land allotted by virtue of this act shall be inclosed and fenced round with ditches and quickset hedges with proper posts, rails and other guard fences to such quickset hedges – and the said quickset hedges, ditches and fences when properly made shall thereafter be kept up, maintained, scoured and supported by the person or persons whom the same plot shall be allotted'. 'Quickset' hedges are hawthorn hedges, and

Typical Georgian enclosure landscape

hawthorn was the most frequently stipulated shrub. Such was the demand for hawthorns that several nursery suppliers made a small fortune in the nineteenth century supplying little else! It is often claimed, contrary to all the evidence, that the enclosed and hedged landscape of England is wholly Georgian in origin. Dramatic as the impact of the eighteenth- and nineteenth-century enclosures must have been, they were but the most recent episode in the thousand-year story of hedge history. Nevertheless, by the end of the period England and Wales possessed about half a million miles of hedge-row, and the countryside remained unchanged, as our early Victorian ancestors knew it, for the next hundred years.

The word 'hedge' is itself derived from the Anglo-Saxon. The Old English words *hege* ('hedge'), *heg* ('hay'), *haga* and *gehaeg* ('enclosure') seem to have been partly interchangeable. Together with their various case forms and the Old German *hagan* ('hedge'), they have found their way into numerous place names such as Haw, Hawcoat, Haydon, Hayes, Hayne, Haynes, Haywards Heath and Haywood. Thus 'hawthorn' too means 'hedgethorn'. The modern French word for hedge is *'haie'* and the German *'hecke'*. A number of plants and animals are also eponymously associated with the hedgerow; hedgehog, hedge sparrow, the hedge brown butterfly and hedge parsley immediately spring to mind. However, the list is much increased when traditional local names are considered as well. Geoffrey Grigson, in *The Englishman's Flora*, lists more than forty 'hedge' plant names. These include the delightful 'hedgy-pedgies' from Wiltshire for the dog-rose, 'hedge lovers' from Devon for herb Robert and 'hedge bells' for bindweed from the south of England.

It has been established that about 70 per cent of the British Isles was once covered by forest – that is, most of the land below about 610 metres, except for those parts of the North and West where the climate is so moist that peat bogs represent the natural vegetation rather than woodland. Today, after Iceland, Ireland and the Netherlands, we are the least wooded nation in Europe, with only about 9 per cent of the country wooded. However, as most of the country was originally woodland, it naturally follows that much of our wild life consists of woodland species.

However, it would be wrong to imagine that our pre-agricultural landscape consisted of one enormous, uninterrupted sea of forest. Dead and fallen trees create gaps which attract the larger mammals such as deer, whose browsing and grazing activities tend to extend and perpetuate the life of these natural forest clearings. For similar reasons clearings also develop along the borders of rivers and streams and around ponds and lakes. The forest itself would have thinned out to scrub on steep slopes, waterlogged ground and in exposed coastal situations. Thus the forest offers three main kinds of habitat: areas of continuous tree cover, natural forest clearings and the intermediate woodland edge between the two. This last is often especially rich in wildlife, as it not only possesses species of plants and animals which are particularly associated with intermediate habitats but, in addition, it also harbours species more characteristic of the habitats which lie on either side of it. Thus the woodland edge possesses woodland, clearing and woodland-edge species. From an ecological point of view hedgerows can be regarded as linear strips of woodland edge in an otherwise largely unwooded landscape. This accounts for the astonishing variety of the hedgerow wildlife. Many of the most familiar flowers, birds and insects of the hedgerow really belong to forest clearings and the forest edge. Not only are hedges valuable as the only habitat for woodland plants and animals over large areas of countryside, but it is also suggested that they can act as highways along which animals move from one area of woodland to another. In addition, the hedge ditch and bank often represent strips of marsh and old pasture; two other habitats that are rapidly disappearing.

Unhappily, the pressures of modern farming have resulted in an enormous loss of hedgerows over the last thirty years, and we assess the significance of this in greater depth in the final chapter. We begin at the height of spring, in April.

April

And fairey month of waking mirth
From whom our joys ensue
Thou early gladder of the earth
Thrice welcom here anew
With thee the bud unfolds to leaves
The grass greens on the lea
And flowers their tender boon receives
To bloom and smile with thee

John Clare, 'The Shepherd's Calendar'

1 *Primrose*
2 *Garlic Mustard*
3 *Cuckooflower*
4 *Blackthorn*
5 *Lesser Celandine*
6 *Ground-Ivy*

Capricious April opens the curtain on the growing season: off to a characteristic start with April Fool's Day, the month is renowned for its unpredictable showers, its rainbows, its bright moods and sudden changes.

If it thunders on All Fools' Day
It brings good crops of corn and hay.

Proverbially the weather in April has always been important to the farmer, hence such sayings as 'April wet, good wheat', and 'April and May are the keys to the year'. And of course 'April showers bring forth May flowers': this is the magic of April – celandines and primroses, violets and dandelions appear like old friends together with lady's smock, barren strawberry and Jack-by-the-hedge. Bare winter skeletons of tree and hedgerow start to fill out with green again, leaving the cold season behind. Pale green light filters through translucent young leaves, and blackthorn branches, already covered with delicate and glorious white blossom, are beginning to come into leaf. As the month progresses, the country-side is transformed unrecognisably from its winter bleakness: there is the promise of warmth and growth and fulfilment as the

hedgerows become cloaked with green, and busy with wild life once more.

The first butterflies of the year make a hesitant appearance, woken from hibernation by the increasing temperature. Very few of our butterflies spend the winter as adults, but peacocks, small tortoiseshells, commas and brimstones may all be seen flitting along the hedgerow by the end of the month; the latter especially where buckthorn, the caterpillar's food plant, grows. By now both honey bees and bumble bees are busy gathering pollen and nectar from the early spring flowers to fuel new or waking colonies.

By the third week nesting is in full swing; hedge sparrows, wrens, robins, yellowhammers, linnets and chaffinches should all have clutches by the end of the month, whilst in a mild spring thrushes and blackbirds may already be feeding hungry mouths. Towards the end of the month the throaty song of the whitethroat, newly arrived from his winter quarters in Africa, finally closes the door on the winter past.

April, whose name derives from the Latin *aprire*, 'to open', is often the month of Easter, the festival of hope and resurrection which is mirrored in nature at this time of the year. April 23 is not only Shakespeare's birthday, but also, appropriately, St George's Day – our patron saint, who acquired legendary fame when he slew the dragon near Silena in Libya, to save the king's daughter from being eaten by the evil monster, and so became the symbol of the triumph of nobility over wrong and waste.

Oh, to be in England,
Now that April's there,
And whoever wakes in England
Sees, some morning, unaware,
That the lowest boughs and the brushwood sheaf
Round the elm-tree bole are in tiny leaf,
While the chaffinch sings on the orchard bough
In England – now!

Robert Browning, 'Home Thoughts, from Abroad'

The Sussex hedge and the Essex hedge

The two hedges which we are following through the seasons belong to quite different landscapes and are totally different in both character and appearance. It is not possible to appreciate and understand their respective pasts, traditions and wild life without knowing something of the background geography and history that have conspired to shape their individual destinies.

The Sussex hedge is set amid the most heavily wooded country-side in Britain. The High Weald of Kent and Sussex can still give the appearance of an almost unbroken expanse of forest when viewed from the summits of the encircling chalk hills of the North and South Downs. On closer acquaintance it resolves itself into a complex and intimate pattern of small fields, woodlands and steep-sided stream valleys known locally as 'ghylls', with the heathy slopes of Ashdown Forest dominating the higher central ridges. The hedge runs almost exactly from east to west separating the parishes of Rotherfield to the north and Hadlow Down to the south. Peter Brandon, in *Sussex*, regards this part of the High Weald as being an almost unspoilt example of thirteenth-century landscape. The map is scattered with farms and woods with names suggestive of early mediaeval forest clearance. The pattern of clearance was distinctive in this part of Sussex: small fields cut out of the waste, separated by the irregular, narrow strips of woodland called 'shaws'. Our hedge also forms the boundary between two farms, Pinehurst Farm to the north and Huggett's Furnace to the south, so it must have played a significant part through the centuries both as a political boundary and a farm fence.

The wild life of both hedges will be greatly influenced by the soils in which they grow. The soils of the Central Weald present a complex mosaic of clays and sandstones, but our hedge runs unequivocally along a strip of the Ashdown Sand, which has weathered to produce a friable, well-drained, acid, sandy loam. As we shall see, this has stamped its imprint in no uncertain manner on the natural history of the hedge. The map opposite shows that both the hedge and the parish boundary join a small wooded stream in the south-east corner of Paul's Field. This is one of the headwaters of the infant Ouse that eventually finds its way to the sea at Newhaven. The hedge itself presents a tall, rather unmanaged appearance, and is distinctly wider than most people's conception of a hedge. However, as we have seen, in this part of the country there is sometimes a rather fine distinction between a wide hedge and a narrow wood.

In stark contrast, the Essex hedge is set in the middle of the fertile, intensively cultivated area of north Essex in the parish of Pebmarsh, close to the Suffolk border. Compared to the High Weald, woodland in this part of the country is a comparative rarity, and the deep, wooded ghylls and small irregular fields are replaced by the wide open spaces of the country's prime cereal-growing region. Thousands of years ago, the retreating glaciers of the Pleistocene ice-sheets left behind a deep mantle of debris covering the under-lying geological deposits of this part of Britain. This chalky clay, known as boulder-clay, covers much of Essex, and gives rise to a strong, loamy soil which is particularly suited to the growing of wheat, barley and beans. The map on the right shows that our hedge forms part of Spoon's Hall Farm and runs roughly north–south. The north end of the hedge joins a footpath that runs along the south side of a little stream that eventually finds its way into the River Colne. The hedge is banked on the west side, bordering the field called Rowley, and is ditched on the east side. It gives the appearance of being more carefully managed than the Sussex hedge, being trimmed to a uniform height of about 1.4 metres, and it is about 1.5 metres wide. Its most distinctive feature are three ancient hedgerow trees, one oak and two crab-apples, that have clearly been carefully nurtured over the years.

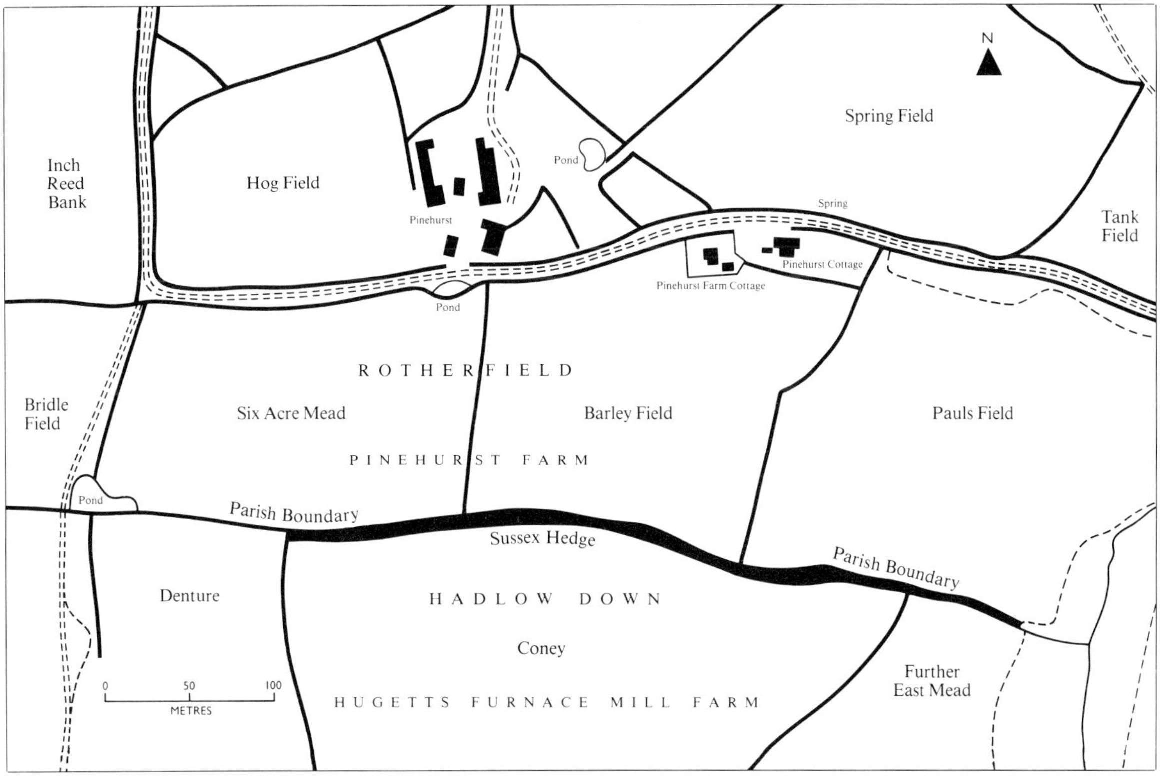

Above
Map of parts of Pinehurst and Huggett's Furnace Mill farms showing the Sussex hedge forming a section of the parish boundary between Hadlow Down and Rotherfield

Below
Map of part of Spoon's Hall Farm in the parish of Pebmarsh, showing position of the Essex hedge

Inch Reed Bank

Hog Field

Pond

Spring Field

Pinehurst

Spring

Tank Field

Pinehurst Farm Cottage

Pinehurst Cottage

ROTHERFIELD

Bridle Field

Six Acre Mead

Barley Field

Pauls Field

PINEHURST FARM

Pond

Pond

Parish Boundary

Sussex Hedge

Parish Boundary

Denture

HADLOW DOWN

Coney

Further East Mead

HUGETTS FURNACE MILL FARM

0 50 100
METRES

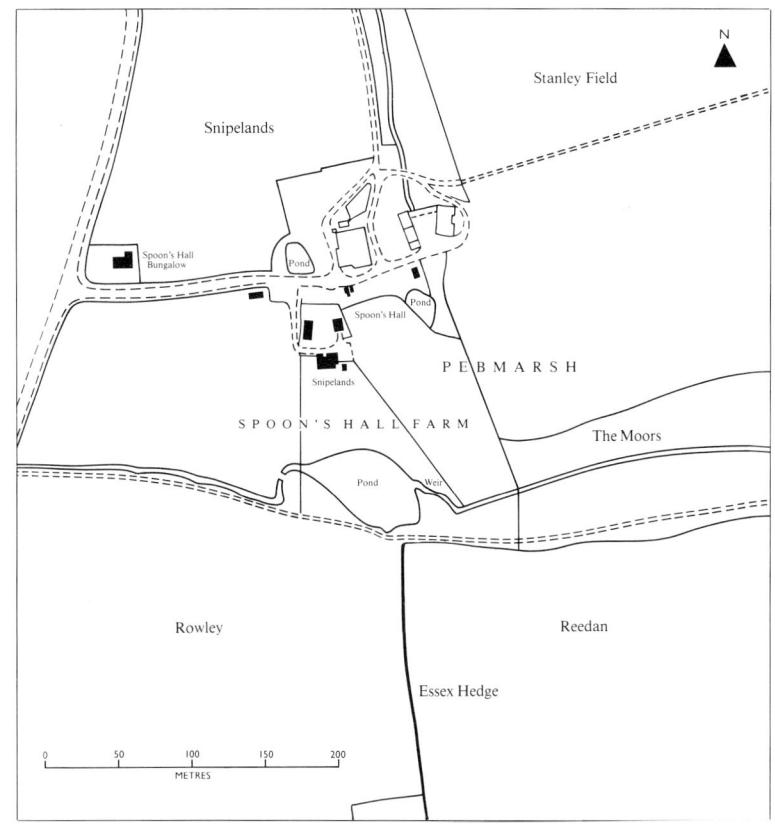

Stanley Field

Snipelands

Spoon's Hall Bungalow

Pond

Pond

Spoon's Hall

PEBMARSH

The Moors

Snipelands

SPOON'S HALL FARM

Pond

Weir

Rowley

Reedan

Essex Hedge

0 50 100 150 200
METRES

21

Trees and shrubs

The real character of a hedge rests in the trees and shrubs which make up its fabric. The particular species found in a hedge depend on a number of different considerations. These include its history, the way in which it is managed, its distance from the nearest patch of woodland, the kind of soil and the part of the country in which it is growing. Altogether, we found twenty-three different trees, shrubs and woody climbers in our two hedges, which is a pretty high proportion of all the hedgerow shrubs that occur in the country. The total list is set out in the table on the right. The interesting thing is that only thirteen species occur in both hedges, and it is intriguing to speculate why this should be. It is to be expected that some of these species will be found in almost all hedges everywhere.

The most frequent hedgerow shrub is undoubtedly the hawthorn, as from the earliest days of the Parliamentary enclosures onwards all new farm hedges were planted with it to the almost total exclusion of everything else. Dog-rose and bramble are also almost universal, whilst the fourth most widespread species is probably the sloe or blackthorn. The sloe first comes into flower in March, and is now in full bloom in both of our hedges. The flowers actually come out before the leaves appear, and in a good year a sloe hedge can appear as if covered by snow. There is a Surrey proverb, 'It is always cold when the blackthorn comes into flower'. In the spring, country wisdom offers this advice:

When the sloe tree is as white as a sheet,
Sow your barley whether it be dry or wet.

In a few parts of the Midlands and eastern England a closely related species from south-east Europe and western Asia has become established, the cherry-plum. It comes into flower even earlier than the sloe, from which it can be told by the leaves and flowers, which both come out together, and by its glossy green twigs. It will be noticed that all of these belong to the rose family, or Rosaceae, and furthermore all are armed with powerful spines and prickles which help to create the ideal, impenetrable, live, stock-proof fence. Also it is these same species that provide the rich autumn hedgerow harvest of hips, haws, blackberries and sloes.

When we look at the other species on the list we begin to learn a great deal more about our hedges. For instance, the presence of both hazel and maple suggests that the hedges are probably quite old, a point that we shall be taking up later in the year. Indeed, hazel is probably the commonest shrub in most older hedgerows, and occupies a particular place in our affections as, throughout the countryside, the first whisper of spring is signalled by the yellow lambs'-tails as they wag in the February wind. These catkins are actually long tassels of male flowers, the yellow colour being the pollen in the ripe stamens. Many plants that flower early in the year depend on wind pollination, as few insects are then

Trees and shrubs of the Essex and Sussex hedges

	Essex	Sussex
Oak	*	*
Beech		*
Ash	*	*
Wild Cherry		*
Blackthorn	*	*
Common Hawthorn	*	*
Midland Hawthorn	*	*
Crab-apple	*	
Dog-rose	*	*
Hazel	*	*
Maple	*	*
Elder	*	*
Hornbeam	*	*
Sallow		*
Holly	*	*
Yew		*
Bullace		*
Dogwood	*	
Bramble	*	*
Ivy	*	*
Elm	*	
Traveller's Joy	*	
Spindle		*

about. However, in order to compensate for the absence of a convenient insect postman to transmit the pollen from the male flowers of one plant direct to the female flowers of another, they have to produce huge amounts of pollen so as to ensure that at least one pollen grain gets blown by chance to the receptive stigmas of the female flowers. Indeed, it has been estimated that each hazel catkin produces almost four million pollen grains! If the catkins are strings of male flowers, where then are the female flowers? If the twigs of the hazel are examined carefully at the time when the pollen is being released, some of the buds will be found to be crowned with a little crimson tuft. These are the female flowers, and the crimson tufts are the receptive stigmas. It is difficult to appreciate that after pollination these minute flowers will develop into next autumn's familiar hazel nuts or 'cobs'.

The field maple is the only species of maple native to Britain, the sycamore having been introduced from Asia Minor in the seventeenth century. It is common in both of our hedges, and, as well as telling us that they are likely to be old hedges, it also suggests that they are situated somewhere in the lowlands, as the maple does not get as far north as Scotland and is rare in north-east England, the extreme South-west and west Wales.

Another shrub that occurs in both hedges is the woodland or Midland hawthorn, and this has an even more explicit distribution, occurring only in the Midland counties south of a line from the Humber to the Severn and in the South-east.

Elder is a shrub that requires a fertile soil, and usually turns up where a gap has formed in the hedge and the soil has been disturbed or enriched by animals and their droppings; this is the reason why it may often be found growing around rabbit burrows.

Many farm hedges possess large standard trees, spaced more or less regularly, and of these the oak is by far the commonest, often showing evidence of having been regularly pollarded in the past. The acorns seed themselves readily into the hedgerow, the young saplings growing up in the protection of the surrounding shrubs. Because of their value to the farmer, these hedgerow oaks were always carefully preserved and managed.

Another common hedgerow tree over the whole country is ash, and the wood was also much valued by the farmer. It does tend to be more frequent on the richer, more fertile soils.

Hornbeam is present in both hedges, and there is a very fine tree between the farm end of the Essex hedge and the stream. Hornbeam has a superficial resemblance to beech, from which it can be told by the toothed edge to the leaves. Although widely planted all over Britain, it is probably native only in the south-east of the country.

There are more species growing in the Sussex than the Essex hedge; nineteen against seventeen. This may simply be due to the fact that the Sussex hedge is rather longer than the Essex one, or it may be due to other factors that we shall uncover later in the year. However, it is worth noting at this point that there are four species that occur in the Essex hedge but not in the Sussex. These are

23

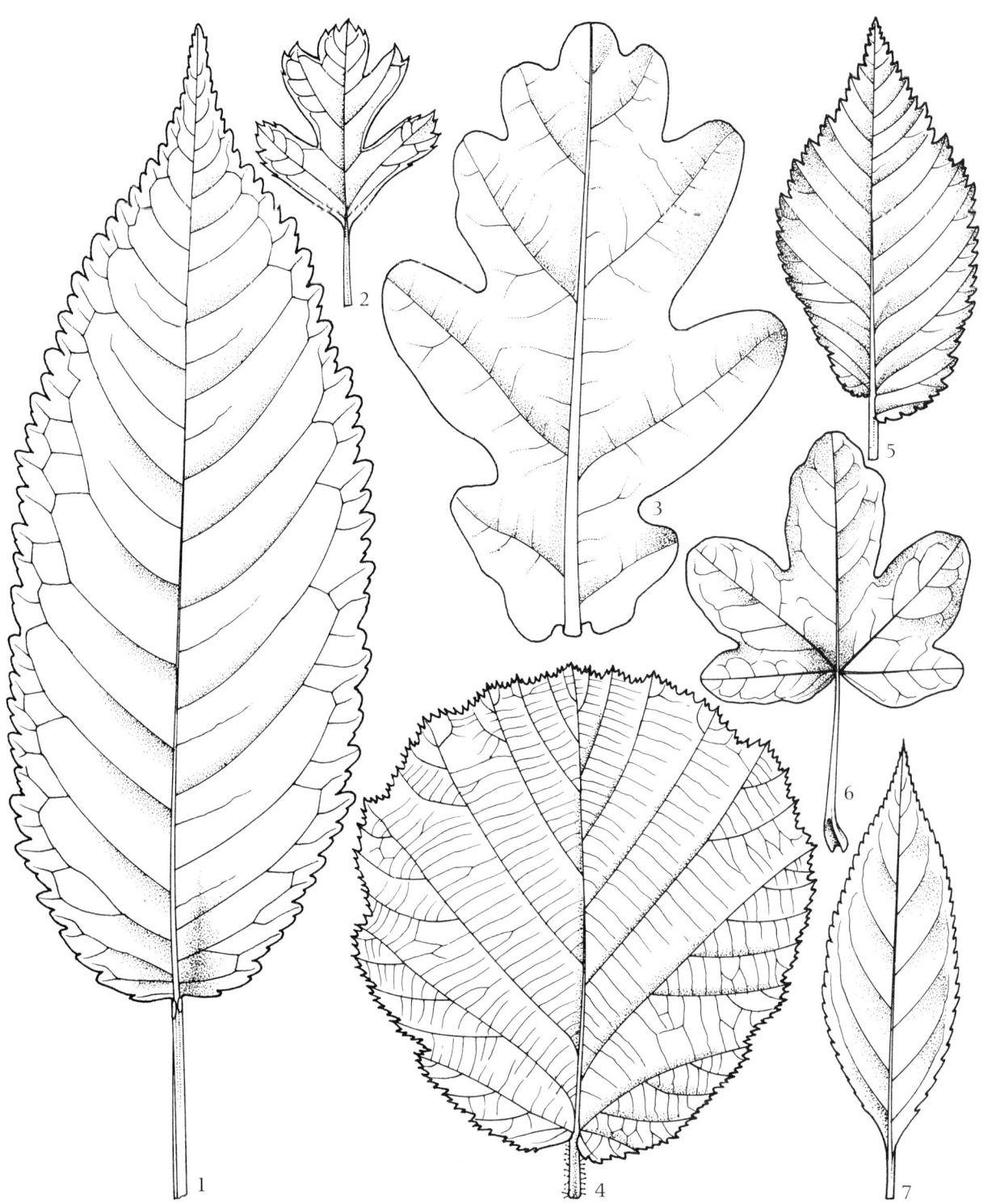

Leaves of some common hedgerow trees and shrubs

1 *Wild cherry* Prunus avium
2 *Common hawthorn*
  Crataegus monogyna

3 *Pedunculate oak* Quercus robur
4 *Hazel* Corylus avellana

5 *Smooth-leaved elm* Ulmus carpinifo
6 *Field maple* Acer campestre
7 *Spindle* Euonymus europaeus

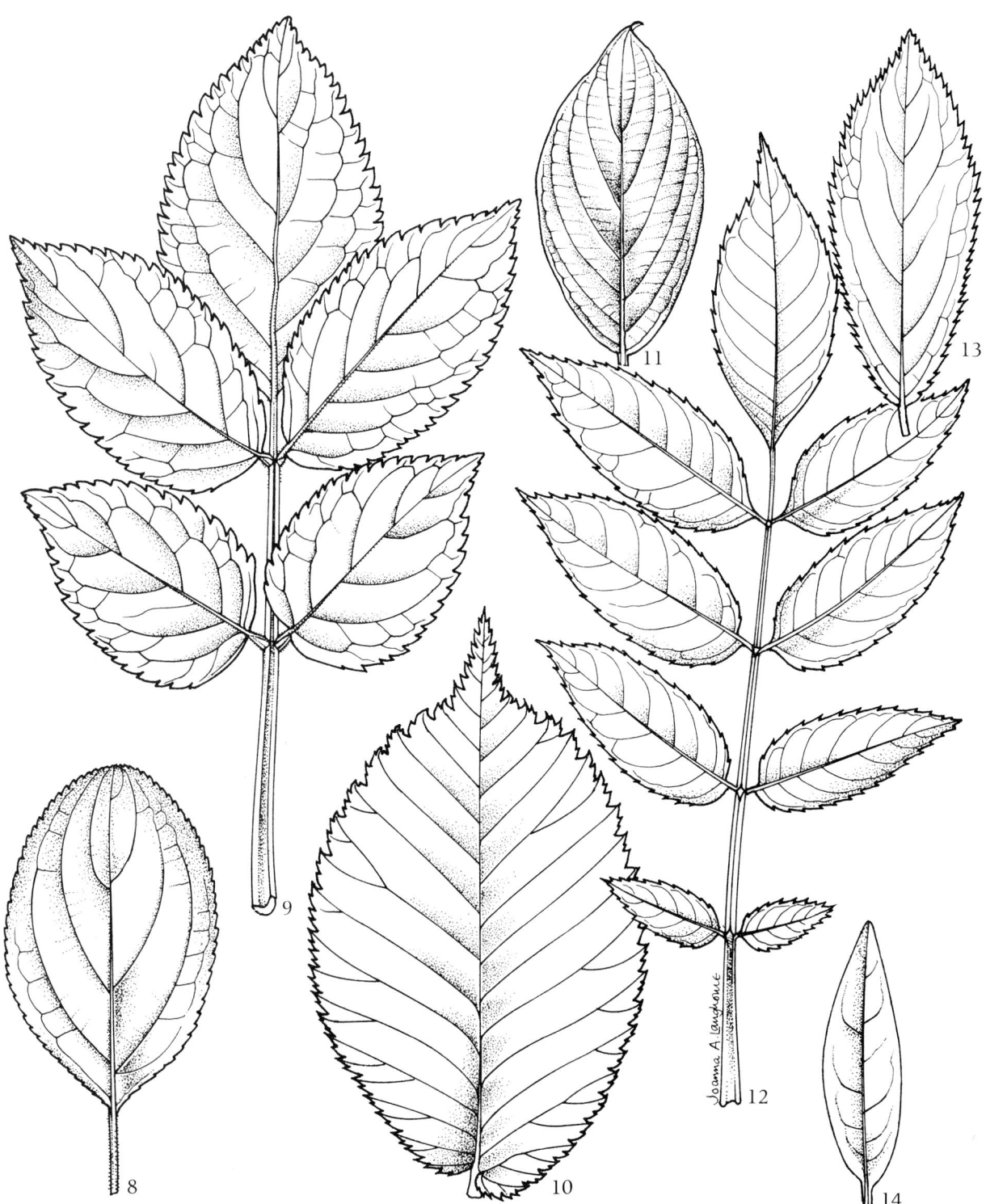

8 *Buckthorn* Rhamnus catharticus 10 *English elm* Ulmus procera 12 *Ash* Fraxinus excelsior
9 *Elder* Sambucus nigra 11 *Dogwood* Cornus sanguinea 13 *Blackthorn* Prunus spinosa
14 *Privet* Ligustrum vulgare

smooth-leaved elm, crab-apple, dogwood and wild clematis or traveller's joy. Smooth-leaved elm is a far more characteristic tree of East Anglia than the Weald, whereas the two large crab-apple trees in the Essex hedge might have been deliberately planted. Dogwood and clematis, on the other hand, tell us that the soil is in all probability lime-rich, rendering it neutral or slightly alkaline in reaction rather than acid, and this ties in very nicely with what we know about the boulder-clay soils of the Essex farm. Neither species would be expected on the acid soils of the Weald, and both are absent from northern Britain, probably because of the cooler summer temperatures.

There are six species present in the Sussex hedge that are absent from the Essex: beech, cherry, sallow, yew, bullace and spindle. Contrary to popular opinion, neither the beech nor the yew is confined to chalk or limestone On the other hand, both have a definite preference for well-drained soils, so that their presence in the sandy Sussex hedge bank is not unexpected, whilst the heavier clays of the Essex farm would be less to their liking. The foliage of yew is very poisonous to stock, so that not surprisingly it is kept cut well back in the Sussex hedge. The wild cherry or gean is a beautiful tree at this time of year, being covered with a mass of white blossom. It is probably the ancestor of the sweet cultivated cherries in the same way as the bullace is usually regarded as the ancestor of the damson. The sallow is a more frequent hedgerow shrub in wetter soils, and it has no doubt spread up from the stream-side into the Sussex hedge.

Our hedgerow trees and shrubs are not difficult to identify, and drawings of the leaves of the more common species can be found on pages 24 and 25, including some like privet and buckthorn which do not happen to occur in either of our two hedges.

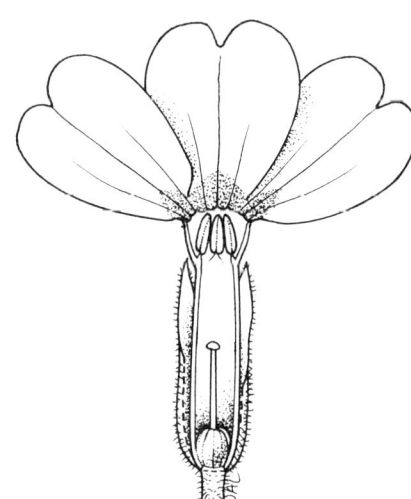

Primrose – thrum-eyed

Spring flowers

A hedgerow bank bright with primroses, violets and barren straw-berry must be one of the most evocative sights of spring. Indeed, so indissolubly linked are spring and primroses in our subconscious that it comes as a shock to realise that in an increasing number of areas spring-time no longer comes with primroses. Fortunately, the uprooting of wild plants in the countryside is now prohibited by law.

Among thy woodlands shady nooks
The primrose wanly comes.

John Clare, 'The Shepherd's Calendar'

Primroses come in two forms, the 'thrum-eyed' and the 'pin-eyed'. The five petals are united at their base into a long, narrow tube, and in the pin-eyed flowers the stamens are inserted about halfway up the tube, the long style reaching to the top of the tube with the stigma showing like a pin-head in the middle of the flower. In

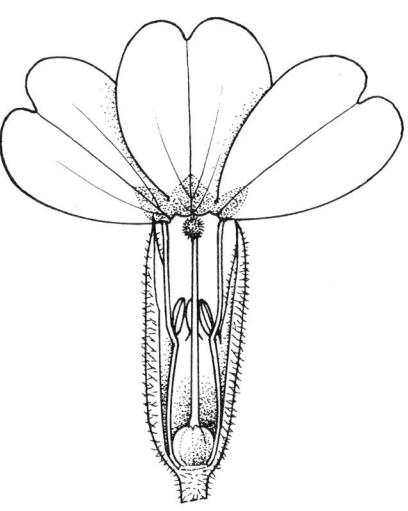

Primrose – pin-eyed

thrum-eyed flowers the five stamens are attached close to the top of the tube whilst the style only reaches halfway. This unusual arrangement is shown on the left and its purpose is to reduce the chance of self-pollination or, more specifically, to prevent a flower being pollinated by pollen from another flower on the same plant. Charles Darwin described his observations on the effects of this arrangement in 1877. He found that pollination of a 'pin' flower by 'thrum' pollen or vice versa resulted in a much higher degree of fertilisation than pollination by pollen from a flower of the same kind. The reason is that 'thrum' flowers have larger pollen grains than 'pin' flowers. Likewise the smaller pollen of 'pin' flowers will be successful only on the stigmas of 'thrum' flowers. If the primroses along a hedgebank are counted, it will be found that there are approximately equal numbers of the two kinds of plant. Darwin also noticed that apparently few insects visit the flowers, and he suggested that night-flying moths might be the chief pollinators. Bees and bee-flies are also occasional visitors, but the identity of the main insect pollinators remains something of a mystery and one that well might be resolved by the patient observer!

In parts of England it is considered unlucky to bring fewer than thirteen primroses into a house when picking the first posy of spring: the old wives' tale goes that if you pick a smaller number, this will be the number of eggs that each hen of the household will hatch during the entire year. If you took one flower indoors, only one chick would hatch out of the entire brood. In other parts of the country this superstition applies to geese. It is quite commonly believed that to bring a single primrose into the house is unlucky as it heralds a death in the family; yet hanging a bunch of primroses in the cowshed will protect it from the antics of witches. Children who eat the flowers are given the power to see fairies. Petals are pulled off one by one for love-divination, and the primrose is said to be lucky for love and marriage. In Germany, where it is called 'key flower', people believed that it could reveal buried treasure and open locks.

Primrose tea, which is prescribed homoeopathically today, is good for rheumatism, arthritis and migraine as well as being a general blood-cleanser and a non-addictive cure for insomnia. The mediaeval herbalists made typically optimistic claims that it could cure all manner of ills, from 'frenzie' to the king's evil. However, the dried roots do contain saponin and, as senegar root, are used as expectorants in modern medicine.

The primrose was the favourite flower of Benjamin Disraeli, Lord Beaconsfield, who always wore one in his buttonhole, and in 1882 Primrose Day was invented; ever since, Disraeli's statue has been decked with primrose wreaths on 19 April. The Primrose League, founded in his honour, sets out to perpetuate his constitutional principles. Now, ironically, primroses are scarce around Beaconsfield due to over-enthusiastic picking.

Dog-violets make traditional company with primroses, and a careful search should discover two different species on the spring hedgebank. The early dog-violet (*Viola reichenbachiana*) starts

flowering three to four weeks earlier than the common dog-violet (*V. riviniana*), but is not found north of the border in Scotland and only infrequently in Wales and the South-west. The easiest way to tell the difference between the two is by the colour of the spur at the back of the flower. In the early dog-violet it is as dark as or darker than the petals, whereas in the common dog-violet it is paler. Hedgebanks on lime-rich or chalky soils may also boast the hairy violet (*V. hirta*) which differs from both the dog-violets in having spreading hairs on the leaf stalks. The hairy violet occurs on the Essex hedgebanks where the chalky boulder-clay provides it with all the lime that it needs, but it is predictably absent from the Sussex hedge.

The delightful little barren strawberry, so teasingly similar to the later flowering, true wild strawberry, is one of the earliest flowers of spring. The rather blue-green leaves should be sufficient to give it away, but if in any doubt the real trick is to look at the terminal tooth of the middle leaflet. In the barren strawberry it is shorter than the two teeth on either side, while in the wild straw-berry it is as long as or longer than its two neighbouring teeth.

In the damper hedgerows the lilac-coloured flowers of the cuckooflower, known in various parts of the country as lady's smock, milkmaids or mayflower, make a lovely show. It belongs to the wallflower family (Cruciferae), and in a few weeks' time will act as the food plant for one of our most characteristic hedgerow butterflies, the orange-tip. It is a flower of the fairies whose magical powers are to be respected: not to be picked and brought into the

Common dog-violet
Viola riviniana

house for fear of its being struck by lightning or of incurring a storm. The flower is used in love-divination by picking off the petals one by one, but was also used by witches in their spells, so if it were found in a May-day garland it was immediately torn to pieces. It is believed that where it grows abundantly the earth is rich in metals.

April and May are 'when the cuckoo doth begin to sing her pleasant notes without stammering' (Gerard), and it is probably called cuckooflower because it flowers around the time when cuckoos are first heard. One quaint explanation for the name 'lady's smock' is that the flowers look like little smocks hung out to dry, as they used to be once a year in the early spring.

Also called meadow bittercress, it is rich in minerals and vitamins and its leaves were often used in salads, being commonly found on market stalls in the old days. It gets its Latin name *Cardamine pratensis*, from two Greek sources, one meaning 'watercress', from its flavour, the other from its supposed medicinal value as a heart sedative. *Pratensis* means 'of the meadows'.

A real hedge plant, the hedge garlic or garlic mustard, commonly known as Jack-by-the-hedge, is so named because of the strong garlic smell given off by the crushed leaves. It has been used in sauces for centuries – traditionally to go with salt fish or boiled mutton – hence another name, sauce alone. Its small, white, star-like flowers are among the first to bloom along the hedgerows. Like the cuckoo flower, it is a member of the wallflower family, and it too provides food for the caterpillar of the orange-tip butterfly. It possesses medicinal and antiseptic properties: Gerard says, 'the seeds bruised and boiled in wine is a good remedy for the wind, colic, or the stone, if drunk warm'. One quaint belief holds that sniffing the leaves like snuff will calm someone who is hysterical.

The most blatant of the early spring hedgebank flowers is surely the lesser celandine with its bright glossy, golden yellow, star-shaped flowers. In a mild year they can be out as early as mid-February. A distinct sub-species of the plant can be found in shadier and damper places which develops little buds or bulbils at the base of the leaves which enable it to reproduce vegetatively.

There is a Flower, the lesser Celandine,
That shrinks, like many more, from cold and rain;
And, the first moment that the sun may shine,
Bright as the sun himself, 'tis out again!

William Wordsworth

Golden guineas, starflowers, golden stars do indeed close their flowers in dull weather and open in the sunshine. The name 'celandine' comes from the Greek for a swallow, 'bycause that it beginneth to springe and to flowre at the comming of the swallows, and withereth at their return', says a sixteenth-century herbal. Its generic name, *Ranunculus*, comes from the Latin for a frog, because, like the creature, the flower is to be found in damp habitats.

One of its local names is pile-wort: lesser celandine has little tubers on its root (*ficaria*, its specific name, means 'fig-shaped'), which indicated to herbalists in the past that it was a sure remedy for this complaint because of the similarity between the shapes. Indeed, it has been successfully used to treat haemorrhoids, and juice from the root was once applied to warts. The young leaves, because of their vitamin C content, were eaten to prevent scurvy.

A much more self-effacing plant is the ground ivy as it creeps along the ground in the hedge bottom, with its little scalloped, kidney-shaped leaves and violet flowers like dead-nettles. Several country names proclaim the traditional association of this little plant with the hedge, hedge-maids and Robin-run-in-the-hedge being but two. Its shy habit hides a rather unusual feature. A careful examination will reveal that some plants possess small flowers and others, distinctly larger flowers. The large flowers are hermaphrodite – that is they possess both the female style and stigma and the male stamens, whereas the smaller flowers are female only. This device, which is shared by only a few other British plants, increases the probability of cross-pollination, which is further enhanced by the stamens of the hermaphrodite flowers ripening before the stigmas become receptive.

In the past, ground ivy played an important part in medicine and also in the brewing of beer. Before hops were introduced for this purpose in the sixteenth century, ground ivy was widely used to clarify and flavour ale, and it also improved its keeping qualities. This accounts for local names like alehoof and tunhoof: the juice of the leaves, tunned up in ale, was thought to cure jaundice and other complaints.

As far back as the second century AD Galen was aware of the use of ground ivy for treating inflamed eyes, and its leaves used to be sold on London streets for making a lotion for soothing tired or sore eyes. It was used in lung complaints, and was at one time the medicine of hope for consumptives and was said to relieve asthma. An infusion of the leaves was used on bruises, and this tea, known as gill tea, was a tonic and diuretic, and drunk as a remedy for coughs, indigestion and kidney complaints. Its local name, 'gill over the ground', comes from the French '*guiller*', to ferment ale. The dried leaves were taken as snuff, to relieve headaches.

It is a relative of sage, mint and lavender, and is rich in vitamin C. It is edible, and one traditional English recipe describes pork stuffed with ground ivy leaves. A valuable plant indeed, it is even said to cheer away melancholy!

In the older hedgerows several of the early spring flowers more usually associated with true woodland may be found. In both the Sussex and the Essex hedges one such is the dog's mercury. Hardly a beautiful plant, it is unusual amongst British wild flowers in that it produces separate male and female plants. The male flowers, which are produced on long, pendulous catkins, are minute and greenish in colour, as are also the long-stalked female flowers. Dog's mercury comes into flower as early as February when there are few pollinating insects about, so wind pollination is a sensible

Chickweed Stellaria media
(*see page* 33)

30

Wood anemone
Anemone nemorosa

strategy, especially as the plants usually grow in large clumps or extensive carpets. It has been suggested that female plants are commoner in the middle of woods whereas the male plants prefer the woodland edge. On this basis we should expect hedgerow plants to be more often male, but it is actually rather difficult to confirm such a sex discrimination! It may be the case that the female flowers prefer damper and the male, drier places. When all is said and done, a damp soil is better for seed germination, and dry air for wind pollination.

Another woodland plant that grows in the Sussex hedge is the wood anemone, in most years coming into flower in March. It must occasionally have flowered a lot earlier than this, as in Somerset it is called 'Candlemas caps', Candlemas being 2 February, the feast of the Purification of the Blessed Mary, with whom the plant was often associated. Its alternative name, 'Easter-flower', is more realistic. Also called the 'windflower', it is said to flourish in windy places or to come out when the wind blows.

Legend waxes lyrical about the anemone as a love flower. Mortally wounded by a boar, Adonis lay in the bloodstained grass where he was found by Venus; overcome with grief she swore that her lover should live forever as a flower, and anemones sprang up where her tears fell. So the flower is sacred to Venus and has the power to inspire love.

In some countries people believe the air to be so tainted where wild anemones grew that it would cause severe sickness, and they do have an unpleasant smell. The plant contains proto-anemonin and is in fact poisonous. Its name derives from the Greek *'anemos'*, wind; its specific name, *nemorosa*, means shady – from its habitat.

31

April Recipes

Nettle Quiche
Stuffed Dock Leaves
Chickweed and Cheese Soup
Spring Salad
Birch-shoot and Marmite Sandwiches
Dandelion Wine

Nettle Quiche

Serves 4

225 g/8 oz cooked nettles
25 g/1 oz finely chopped
   shallots or spring onions
25 g/1 oz butter
3 eggs
275 ml/$\frac{1}{2}$ pint cream or milk
1 20-cm/8″ pastry shell,
   baked blind
25 g/1 oz grated
   Parmesan cheese
salt, pepper and nutmeg

Pick the young shoots, the four top leaves, of stinging nettles, wearing rubber gloves to do so! Wash them and cook them in their own water, just as you would for spinach. Add some salt, and simmer for about ten minutes until the leaves are soft. They lose their sting when boiled. Drain them and they are ready for use.

Cook the shallots in the butter for a minute or so. Add the cooked nettles and stir over a moderate heat to evaporate any liquid. Season to taste. Beat the eggs and cream in a bowl until fluffy, season, and stir in the nettles. Pour into the pastry case, sprinkle with the cheese and a little nutmeg, dot with more butter and bake for twenty-five to thirty minutes at 375°F, 190°C or gas mark 5.

Stuffed Dock Leaves

Choose young, small dock leaves and wash them, then dry them on a towel. Make a stuffing with a mixture of cooked rice, grated cheese, chopped creamed spinach or comfrey, and seasonings. Stir over a gentle heat in a little butter until well-mixed, then put a teaspoon of the stuffing on the dock leaf and roll it up. Secure with wooden toothpicks and brush with olive oil, pack in an ovenproof dish and bake at 350°F, 180°C or gas mark 4 for twenty to twenty-five minutes. Serve warm or cold, as an appetiser.

Chickweed and Cheese Soup

Serves 4

1 large bunch of chickweed
40 g/1½ oz butter; 10 g/½ oz flour
425 ml/¾ pint stock; milk
50 g/2 oz grated cheese

Wash and trim the chickweed and cut off all the straggly roots. Chop it finely and simmer it in the melted butter for ten minutes. Stir in the flour, then gradually add the stock, stirring all the time. Liquidise it and add a little top of the milk. Stir in the grated cheese and serve when it has melted.

Spring Salad

young dandelion leaves
hazel shoots; tomato
hedge garlic leaves; vinaigrette

Wash and tear the dandelion leaves, wash the hazel shoots and slice the tomato thinly. Wash and shred the hedge garlic leaves and toss all together in an oily vinaigrette.

Birch-shoot and Marmite Sandwiches

Butter some very thin slices of fresh bread, half of them white and half of them brown. Spread thinly with Marmite and sandwich with little birch shoots, washed and dried. Cut off the crusts and cut into triangles.

Dandelion Wine

(See notes on wine-making, page 104)

2 quarts of flower heads
4 litres/7 pints hot water
225 g/8 oz sultanas
1 lemon
1 sachet wine yeast
1 k 125 g/2½ lb sugar
1 Campden tablet

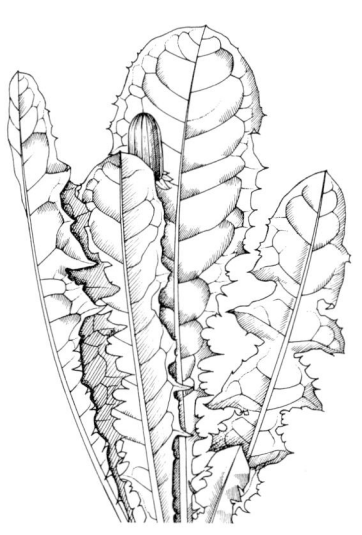

Pick the flowers on a sunny day in late April or early May when the blooms are fully out. Cut off the stalk and green calyx, leaving only the yellow head. Place these in a suitable, sterilised vessel, and pour the hot water over them. Rub the flowers against the side of the vessel with the back of a sterilised wooden spoon to extract the flavour. Cover and leave to cool.

Wash and chop the sultanas, thinly pare the lemon rind and squeeze the juice, discarding all the pith. Add these to the cool flower water together with the yeast. Replace the cover and ferment on the pulp for four days, pressing down the floating flowers and sultanas twice each day.

Strain out, press and discard the solids, stir in the sugar and pour the must into a sterilised fermentation jar. Fit an airlock, and leave in a warm room for about three weeks until the fermentation is finished.

Move the jar to a cold place for a few days to encourage the wine to clear, then siphon it into another sterilised jar and discard the sediment. Top up with cold boiled water, add 1 Campden tablet to prevent infection, bung tightly, and store in a cold place until the wine is crystal clear. Bottle it and keep the wine until Christmas. Serve it chilled as a social wine: it will be slightly sweet.

You can make elderflower and May blossom wine in the same way. Use no more than 1 pint/570 ml elderflowers or 3 pints/1 litre 700 ml May blossom.

May

Come queen of months in company
With all thy merry minstrelsy
The restless cuckoo absent long
And twittering swallows chimney song
With hedge row crickets notes that run
From every bank that fronts the sun
And swathy bees about the grass
That stops with every bloom they pass
And every minute every hour
Keep teazing weeds that wear a flower

John Clare, 'The Shepherd's Calendar'

The Merry Month of May gets off to a festive start with May Day on 1 May, when traditional pagan ceremonies were – and in some places still are – performed, symbolising the turning cycle of the year and ushering in summer. The maypole on the village green, itself a symbol of fertility, was decked with long ribbons, and children danced around it weaving in and out, plaiting and un-plaiting the ribbons. A Queen of the May was chosen from amongst the young girls, and she, along with the May-doll, represented the reincarnation of Flora. In some places a Lord of the May was elected too, and Jack-in-the-Green, representing Spring, performed his ritual dance. The Queen would take a garland or hoop of flowers – sometimes with the May-doll in the middle – to the church, serenaded by musicians and Morris dancers. It was said that fairies and witches would get up to their antics on this day, particularly in the dairy, where milk and butter were likely to be bewitched. In Oxford, every year at 6 o'clock on May Day morning the choristers of Magdalen College greet the sunrise by singing a Latin hymn from the top of the tower, and this is followed by a peal of bells and Morris dancing in the streets.

Various religious festivals fall in May, too – Whitsun, Ascension and Rogation Sundays: on the latter clergy used to go out into the

1 *Alexanders*
2 *Cow Parsley*
3 *Greater Stitchwort*
4 *Red Campion*
5 *Hawthorn*
6 *Lords-and-Ladies*

34

fields to bless the crops. On Rogation days, just before Ascension, the elders of the parish would walk the boundaries with their villagers, and whip the young boys with willow wands at certain points so that they would remember the position of the boundary: where it was marked with a stream they would be ducked. Fortunately for the boys, the coming of the Ordnance Survey has put an end nowadays to the necessity for such rituals.

May 29 used to be known as Oak Apple Day, in remembrance of when King Charles II took shelter in an oak tree at Boscobel in 1651 whilst fleeing from Cromwell after his defeat at the battle of Worcester. After his restoration in 1660 this date became the day when, for many years, May Day was celebrated.

The weather in May is variable, but as often as not it is a wet month. There are many country sayings which indicate this, such as 'A leaky May and a dry June puts all in tune', 'Rain in May makes bread for the whole year' and 'May makes or mars the wheat', as well as

Mist in May, heat in June,
Makes the harvest come right soon.

Moreover, the cold weather is by no means always over:

Who doffs his coat on a winter's day
Will gladly turn it on in May.'

And of course there is the well-known 'Cast not a clout till May be out'. There are two proverbs, one for those who desire immortality, and the other for those hoping for a happy marriage: 'He who would live for aye must eat sage in May'; and 'Marry in May, repent alway'.

May is the month when bluebells carpet the woods, buttercups, the meadows, and the hedgerows are festooned with hawthorn and crab-apple blossom. Grassy banks abound with cow parsley, stitchwort, red campion and wild arum. The list is a long one, but not too long for John Clare:

My wild field catalogue of flowers
Grows in my rhymes as thick as showers
Tedious and long as they may be
To some, they never weary me.

John Clare, 'The Shepherd's Calendar'

Birds

Nesting activity in the hedgerow reaches its peak during May. The importance of hedgerows as bird habitats varies enormously according to the way in which the hedge is managed and in different parts of the country. In many areas of intensive farming, hedgerows are virtually the only cover left for scrub and woodland

1 *Hedge remnant*

2 *Laid hedge*

3 *Mechanically pollarded*

4 *Clipped*

5 *Overgrown, but with no undergrowth*

6 *Overgrown, but stock proof*

7 *As 6, but with bushy outgrowths at the base*

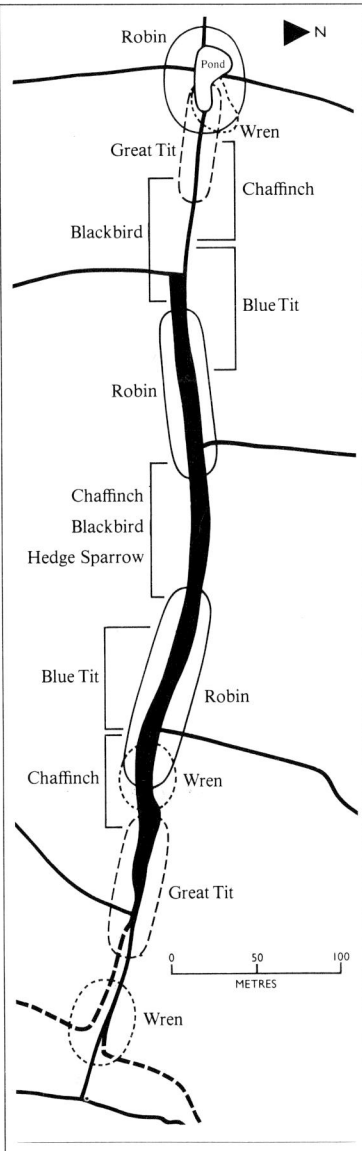

Sussex hedge bird
territories May 1981

Wren	3 pairs
Great tit	2 pairs
Blue tit	2 pairs
Chaffinch	3 pairs
Robin	3 pairs
Blackbird	2 pairs
Hedge sparrow	1 pair
	16

equivalent to 28 pairs/1000 m

nesting species, although in others, like the Weald and parts of Devon, where there is still much woodland, they are less crucial.

In 1967 Professor Norman Moore and his colleagues from Monks Wood Experimental Station carried out a survey of the bird population of hawthorn and elm hedges in parts of Huntingdonshire. For the purposes of the survey they divided the hedgerows into seven different categories, depending upon the way in which they had been managed. These are illustrated opposite.

They found that hawthorn hedges consistently supported more pairs and also more species of birds than elm hedges. Hawthorn produces a much denser hedge than elm, and also comes into leaf earlier, so affording a great deal more protection for nesting. In addition, hawthorn is a much richer source of insect food than elm. The form of management had an even more pronounced effect on the bird population. By far the best was the dense, overgrown hedge with much bushy outgrowth and cover at the base. Hawthorn hedges of this kind supported an average of thirty-four pairs and nineteen different species per 1,000 yards, whereas, as expected, the least good were the hedge remnants with fewer than five pairs per 1,000 yards. There was not a great deal to choose between the other four groups. Of the three various kinds subjected to regular management the clipped hedge, which produces the densest cover, supported the largest number of pairs, whilst the mechanically pollarded produced the largest number of species: ten per 1,000 yards.

The easiest way of estimating the bird population of a hedge is to count the singing males in the breeding season. Almost all our land birds are strongly territorial, and the males proclaim their ownership by singing from a conspicuous vantage point. Birds that nest in open country, like the skylark, where there are no convenient elevated perches, have overcome the problem by evolving the ability to sing as they hover high overhead. The 'dawn chorus' is at its best during the first half of May, the birds becoming more silent as the summer progresses. You can count the singing males by walking along the hedge early in the morning and marking the position of the various individuals on a large-scale map. Take care not to record the same bird more than once, especially as individuals will move about their territory singing from different vantage points. For the greatest accuracy it is recommended that you should carry out the procedure at least eight times at fortnightly intervals between the end of April and July. The maps produced on each occasion can then be collated to produce a reliable picture of the bird population of that particular hedgerow.

The result of our own survey of the birds of the Sussex hedge is shown on the left. We estimated a population of sixteen pairs and seven species; three pairs of wrens, two pairs of great tits, two pairs of blue tits, three pairs of chaffinches, three pairs of robins, two pairs of blackbirds and one pair of hedge sparrows. As the hedge is 634 yards (580 metres) long, these results are equivalent to twenty-five pairs and eleven species in 1,000 yards, or twenty-eight pairs and twelve species in 1,000 metres. The Sussex hedge would

39

probably fall into Professor Moore's 'overgrown hedge with outgrowths' class, so it does not appear to be as rich as his Huntingdonshire hawthorn hedges of the same kind. The reason is probably that our hedge is not so very bushy and dense on its warm southern side. The list is interesting as it is a bit short on the species typical of a normal managed field-boundary hedge. There is only one pair of hedge sparrows and no whitethroats, linnets or yellowhammers, although there is a pair of yellowhammers in the pollarded hedge between Coney Field and Further East Mead. Both the robins and the wrens would be nesting in the hedgebank, an ideal site for the beautiful domed nest of the wren and with plenty of shaded ledges and overhangs for the robins. Only hedges with many mature trees are suitable for the hole-nesting tits, and the two pairs of blue tits and two pairs of great tits do suggest that, so far as the birds are concerned, the Sussex hedge is more like a narrow piece of woodland than a narrow strip of scrub. Both blackbirds and chaffinches are equally at home in hedgerow and woodland, and indeed are probably the two most abundant British birds after the wren.

In their book *Hedges*, Pollard, Hooper and Moore list a total of thirty-four species of birds that commonly nest in hedgerows. In addition to those that we have already mentioned, the upper branches of the mature trees may support the nests of the odd carrion crow or magpie, while holes in the trunks may be occupied by starlings, stock doves and the occasional little owl in addition to the tits. The largest number of species nest in the various hedgerow shrubs, and include the song-thrush, bullfinch, greenfinch, gold-finch, long-tailed tit and lesser whitethroat. The long-tailed tit builds what must surely be the most exquisite nest of any British bird. It is a marvellous soft, domed ball, secreted deep in a thorn bush and consisting of a wonderful concoction of cobwebs, hair, moss and lichens, thickly lined with feathers.

Strangely enough, very few summer visitors are habitually hedgerow species. The two exceptions are the whitethroat and the lesser whitethroat. The whitethroat is one of our commonest species of rough scrub-land and has successfully adopted the hedgerow as an additional habitat. Its numbers catastrophically declined in the early 1970s as a result of a series of droughts in the Sahel region of Africa on the southern edge of the Sahara, which reduced its winter insect food sources, but happily it is now recovering its former abundance. The male's defiant chuckling, throaty song is usually first heard towards the end of April and is a welcome sign of the final passing of winter. The nest is built low down among rank herbage, and this habit no doubt was the reason that earned it the local name of 'nettle-creeper'. The alternative local name of 'hay-jack' is also interesting as it is almost certainly the same as 'hedge-jack'. The lesser whitethroat is more restricted in its distribution, being largely absent from the extreme South-west, from much of Wales and northern England and Scotland. A sleeker bird than the whitethroat, it has a totally different song and usually nests higher up in the dense cover of hawthorn or sloe.

Above
Blackbird
Top
Wren

The most typical of all hedgerow birds must be the hedge sparrow, and its association with hedges is as old as hedges themselves. Its name in Old English was *sucga* or *succa*, but it also appears as *haegsucca* and *hegesugge*, 'hedge sparrow'. It is celebrated in several place names such as Suckley in Worcestershire, Sugwas in Herefordshire and Sugworth in Berkshire. In addition it has acquired a whole variety of local hedge names in different parts of the country: hedge-betty; hedge-mike; hedge-pick; hedge-pip and hedge-spick. Many ornithologists insist on referring to it by its alternative name of 'dunnock', but there really seems little virtue in abandoning the familiar 'hedge sparrow' with its impeccable pedigree merely to demonstrate that the user knows that it is not related to the house sparrow. The sight of a clutch of the sky blue eggs set in their little nest of moss and twigs, hidden deep in the thickest part of the hedge, is one of the delights of the countryside.

Banks and ditches

Hedgerows have been significant features of our landscape for hundreds of years. Many formed the line of political boundaries, and many parish boundaries are still marked by ancient hedgerows today. They undoubtedly played an important part in the social history of the countryside. Many different kinds of evidence can be adduced to build up a picture of the history of a particular hedgerow, and when considered together they can sometimes produce fascinating insights into the changing pattern of the social and physical landscape with startling clarity. Hedges not only vary characteristically in different parts of the country, but patterns and features have also altered over the centuries. In this context the hedgebank and ditch should be regarded as any other archaeological monument, and the detailed recording of its physical structure is an essential ingredient in the construction of its biography.

An accurate survey of the hedge profile does not necessarily require professional surveying equipment – level, levelling staff and range poles – if these are not readily available. Much can be done by anyone with little more than a ball of string and a spirit level, providing that certain basic procedures are followed. A representative section of the hedge to be surveyed is first selected. The technique then involves stretching a length of string at right angles to the line of the hedge between two firm posts at a suitable distance above the ground – say, no more than 30–50 cm above the highest point. The string is then tightened and levelled with the spirit level. The vertical distance from the string to the ground is then measured at regular intervals – say, every 50 cm – along the length of the cross-section. It is most important to ensure that the string remains taut and does not sag. The best way of achieving that is to place the posts at intervals of no more than 2–3 metres apart. If the level of the string is altered from one section to another, the difference in level must be carefully noted.

When all the recordings have been completed the hedge profile can be drawn out to scale; this is most easily done on graph paper.

Choose a vertical and horizontal scale as similar as possible in order to reduce the vertical exaggeration of the profile. Next, draw lightly in pencil a line to the appropriate horizontal scale to represent the string. Then draw in the vertical distances from the string to the ground to the appropriate vertical scale, again in pencil. The ends of these perpendiculars can then be joined producing a scale representation of the profile of the hedge. The basic profile can be improved by recording the various shrubs occurring on the line of the profile and noting both the species and their height. This information can be incorporated in the final plan by using different symbols for the different shrub species recorded. Although the results are obviously not as accurate as would be obtained by using professional surveying equipment, providing that the various procedures are carefully carried out, very satisfactory scale profiles of hedges, ditches and banks can be obtained by this simple method. The documentation and recording of features like hedges is becoming increasingly urgent as more disappear in the face of urbanisation and agricultural improvement, and much is still to be learnt from comparing hedge profiles from different parts of the country and with different documented histories.

The profiles of both the Sussex and the Essex hedges are illustrated together on the right. The first obvious point of difference is that the Sussex hedge is much wider than the Essex hedge, 8·5 m against 1·5 m. As we have already observed, the Sussex hedge, like many others in the Weald, is exceptionally broad, and it would appear from the profile that it has been permitted to spread and develop on its south side to a distance of about 3·5 m from the ditch. The dimensions of the ditch and bank of the two hedges are much the same. The bottom of the Sussex ditch is 66 cm below the present height of the bank, and the bank itself is about 1·75 m wide. There is also evidence that about 30 cm of soil has been lost on the north side of the Sussex bank, presumably as a consequence of down-slope erosion eastwards towards the stream. Both boundaries represent sizeable earthworks, indicating considerable antiquity. Furthermore, the Sussex hedge shows evidence of having been even higher in the past, as the roots of many of the trees and shrubs growing along the top of the bank have become exposed due to soil loss.

One obvious point about the two hedges must be made. That is that in both cases the banks must be older than their respective hedges. There is no means of knowing what period elapsed between the building of the banks and the establishment of the hedges. One might have followed immediately on the other, or the bank could have stood as an unhedged boundary for some considerable time. Neither can we be certain whether the hedge was planted or whether it established itself by natural spread from the surrounding waste, or indeed by a combination of both. What does seem certain is that the roles of the two hedges as boundaries in the past have been quite different. The fact that the Sussex hedge is five and a half times as wide as the Essex hedge, together with the frequency of older trees, both standards and laid, suggests that it

Profile of the Sussex hedge

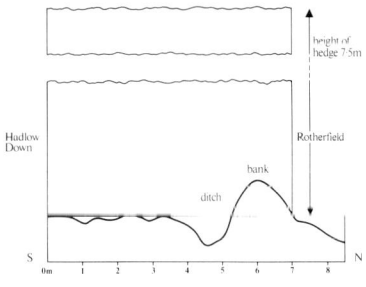

Scale for both hedges
Vertical scale: 1 cm = 20 cm
Horizontal scale: 1 cm = 50 cm
Vertical exaggeration × 2.5

42

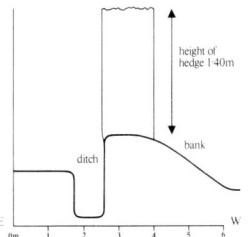

functioned as an important political boundary as well as a simple field fence. Support for this view is found in the fact that it is a parish boundary, and boundaries of most of the parishes in the Weald were well established by the time of the Norman conquest. In contrast, the Essex hedge is today a straightforward, internal farm field boundary. A further insight into some of these questions might be revealed by an archive search for any relevant contemporary documents.

Butterflies

By far the most characteristic hedgerow butterfly of spring and early summer is the orange-tip, so named because the apical third of the forewings of the male is a bright orange. The female, on the other hand, is predominantly white on the upper surface, save for the black tips to the forewings which also have a single black dot in the middle and dark bases. The underside of the hind wings in both sexes is a most beautiful and unusual delicate marbling of green and white, which camouflages the insect perfectly when it hangs at rest in the dappled light beneath the inflorescence of cow parsley. Green coloration is extremely rare in butterflies, and indeed no British species possesses true green pigmentation, the green in the orange-tip being produced by a mixture of black and yellow scales which produce an optical effect of green.

The insect spends the winter as the pupa and usually emerges in May, although in a good season it may be on the wing by the end of April. The eggs are laid in June on the flower or fruit stalks of two very common hedgerow plants, cuckooflower and hedge garlic. They are laid singly, which reduces the chances that the caterpillars will compete for the same food-plant – and indeed, when kept together, the young larvae are frequently cannibalistic. In coloration the caterpillars have an astonishing resemblance to the seed-pods of the food-plant, and when young they possess glandular hairs which exude a sweet fluid that is attractive to ants. In return, the presence of the ants may serve to discourage predators. The larvae feed throughout June and July and finally pupate in August to emerge the following spring.

Another common spring-time hedgerow butterfly is the green-veined white. As its name implies, the veins on the undersides of the wings are bordered with a greenish-grey coloration, which helps to camouflage it when it is at rest among the long grasses of its favourite marshy and damp woodland habitats. Damp hedgerows also suit it very well. The females have two or three dark spots on the upper side of the forewings in addition to the black tip, whereas the male has only one or none at all.

Like the orange-tip, the green-veined white spends the winter as a pupa, and the butterfly emerges at the end of April or the beginning of May. Moreover, the orange-tip and the green-veined white share the same two hedgerow food-plants. However, there the similarity ends, as the green-veined white produces two generations a year. The caterpillars of the first generation feed

43

during June and July, and the adults appear in July and August. These lay eggs, and the resulting larvae feed during August and September and then pupate to emerge the following spring. The adults of the two generations differ slightly in coloration, the second brood being the more strongly marked on the upper side, whereas the pigmentation of the 'green' veins on the underside is more prominent in the spring brood.

Both the orange-tip and the green-veined white can be found throughout the country, and both share the distinction of appearing in the first book on entomology ever to be published in Britain, *Theatrum Insectorum*, edited by Sir Theodore de Mayerne, which appeared in 1634.

A third species of butterfly that may be encountered along the more shaded hedgerows at this time of year is the speckled wood. The winter is again spent as a pupa, and the butterfly emerges at the end of April. It appears again in late summer, as, like the green-veined white, it produces two broods a year.

May flowers

The predominant hedgerow colour in May is white. White hawthorn blossom clothes the hedges, and great white drifts of cow parsley fill the hedgebanks, mixed here and there with the white of the greater stitchwort.

. . . every shepherd tells his tale
Under the hawthorn in the dale.

John Milton, 'L'Allegro'

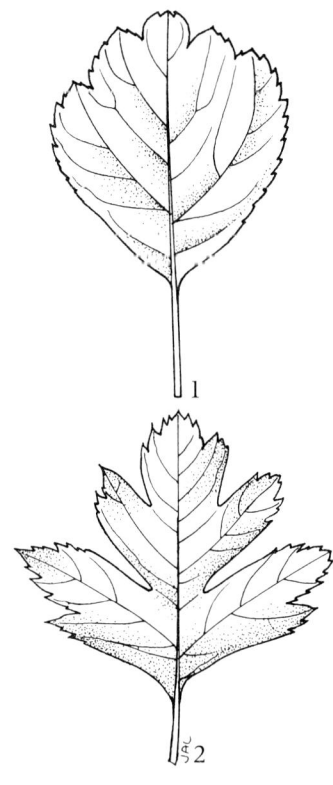

1 *Midland hawthorn*
  Crataegus laevigata
2 *Common hawthorn*
  Crataegus monogyna

We have two species of hawthorn in Britain, the common hawthorn, *Crataegus monogyna,* and the Midland hawthorn, *C. laevigata*. The commonest of the two in hedgerows – and, indeed, probably the commonest of all hedgerow shrubs – is *C. monogyna*. It is usually quite easy to tell one species from the other. The leaves of the common hawthorn are deeply lobed; the lobes are longer than broad, and the sinuses between the lobes are deep, reaching more than halfway to the midrib of the leaf. The flowers have one style, and the berry, 'haw', contains a single stone. In contrast, the leaves of the Midland hawthorn have shallow lobes, these being broader than long and the sinuses between the lobes not reaching as far as halfway to the midrib. The flowers have two styles, and the berry contains two stones. In addition, the Midland hawthorn usually comes into flower about a week earlier than the common hawthorn. The differences between the two species are shown on the right. The picture is complicated by the frequent occurrence of intermediates. This is because the two species will readily hybridise when growing in close proximity, giving rise to fertile hybrids, *Crataegus* × *media*. Furthermore, one or both parents may subsequently disappear, leaving a hybrid population. As its English name implies, the Midland

44

hawthorn has a rather restricted distribution, occurring south and east of a line from the Humber to the Severn, but being absent from large parts of East Anglia and from the South-west and central southern England. Within this area it is much more a shrub of woodlands than the hedgerows. The common hawthorn, on the other hand, is uniformly distributed throughout the whole of the British Isles with the exception of parts of north-west Scotland.

The beautiful blossom of the hawthorn is inextricably linked with the month of May, to the extent that it is known as 'may blossom'. May Day always hinged around hawthorn in the old days, particularly before the calendar changes of 1752 when May Day fell on what is now 12 May, when the hawthorn was in full flower. Essentially it was a festival to welcome summer, and hawthorn blossom was used for decorations, garlands and super-stitious rites. Early in the morning people would go 'a-maying', returning laden with may blossom to decorate their churches and houses, and traditionally they would hang a garland over their front door. King Henry VIII was reported to have gone 'a-maying' from Greenwich to Shooter's Hill 'with his Queen Katherine, ac-companied by many lords and ladies'. There is a Suffolk tradition that the first person who finds a branch of hawthorn in full blossom is rewarded with a dish of cream for breakfast. If you picked the blossom, fasting, on May Day morning, it would protect you against lightning; and girls would bathe their faces in hawthorn dew at first dawn on May Day morning in the hope that it would make them beautiful:

The fair maid who the first of May
Goes to the field at the break of day
And washed in dew from the hawthorn tree
Will ever after handsome be.

Like many of the very supernaturally-endowed plants, hawthorn's magic is equivocal: many people believed it unlucky to bring may blossom indoors because someone will die in the same year. Like-wise it is courting disaster to sleep in a room decorated with the flowers. Never sit under a hawthorn: the fairies will gain power over you. To cut down or destroy a hawthorn is dangerous, and to this day some woodmen refuse to fell it. It is known in some places as 'fairy thorn' where people think it dangerous to gather even a leaf from certain old and solitary trees, because these are the trysting places of the fairies. A 'Scottish Statistical Report' of 1796 says, 'There is a quick thorn of a very antique appearance, for which the people have a superstitious veneration. They have a mortal dread to lop off or cut any part of it, and affirm with a religious horror that some persons who had the temerity to hurt it were afterwards severely punished for their sacrilege.' It is possible that hawthorn acquired the bad side of its reputation from a belief that after the Black Death in the fourteenth century hawthorns sprang up everywhere in the depopulated countryside. It has been suggested that may flowers, which contain trimethylamine – an ingredient of the smell of putrefaction – preserved the stench of London during the Plague.

On the other hand, from earliest times the hawthorn has been regarded as propitious: according to ancient myth it originally sprang from where lightning struck the earth and has been revered as a sacred tree. For the Greeks it was an emblem of hope and a tree of good augury. It was a symbol of marriage and fertility: may blossoms decorated the nuptial altar as well as the wedding festivities, and the bride would wear a wreath of hawthorn. The newly-wed couple were lighted to the bridal bed with torches made of hawthorn wood. A later Christian superstition sprang up that if you picked hawthorn on Maundy Thursday and kept it in the house, it would never be struck by lightning, because

Under a thorn
Our Saviour was born.

It may also have been the Crown of Thorns, and the French call it *'l'épine noble'*. Hawthorn is often depicted as the wreath of the 'Green Man', who is the symbol of returning summer. It is said to have healing properties and to be a protection from lightning:

Creep under the thorn
It will save you from harm.

Hawthorn is a hard and durable wood which in the past was used for making printing blocks and handles. 'Haw' comes from an Anglo-Saxon word meaning hedge, because its sturdy, fast growth and thorny branches have made it an ideal hedging plant for centuries. Hawthorn also has a place in weather lore:

Many haws, many sloes,
Many cold toes.

When the hawthorn bloom too early shows
We shall have still many snows.

The most remarkable of English hawthorns is the Glastonbury Thorn, which is said to have sprung up from the staff of Joseph of Arimathea. The story goes that Joseph and eleven of his followers came to convert the Britons. When preaching to them on Christmas Day at Glastonbury, he struck his staff into the ground as proof of his divine mission, and it immediately burst into life and blossom. A church dedicated to the Virgin was founded on this spot, and the miraculous thorn grew, always blooming on Christmas Day. Sceptics and botanists would call it *C. monogyna var. praecox*.

Hawthorn has been used in medicine as well as magic: its superstitious reputation for healing has grounding in practice, since it provides a non-toxic remedy for such heart conditions as valvular weakness or irregular heart action; it has been used extensively as a remedy for hypertension. The old herbalists

46

prescribed the distilled water of haws for 'any place where thorns and splinters doe abide in the flesh', as it 'will notably draw them out'. But then they had a remedy for everything.

If hawthorn is the commonest hedgerow shrub, then the honour of being the most characteristic hedgerow herb must go to the cow parsley. It is the first of the common wayside umbellifers to come into flower, and its billowing clouds of lacy white flowers act as a great attraction to a host of spring insects seeking nectar and pollen. Hover flies and bees, predatory yellow dung flies and scorpion flies jostle with the orange-tips and green-veined whites. On warm days swarms of black, bristly, long-legged St Mark's flies perform their aerial courtship dance intent on luring the females from their resting places on the cow parsley flowers below.

Pretty though it is, cow parsley is none the less connected with the Devil – perhaps because it can be mistaken for fool's parsley and hemlock, both of which are poisonous. It has been derided as 'devil's meat', 'badman's oatmeal', 'dog parsley' and 'sheep's parsley': evidently providing salading fit only for cattle. Or was it perhaps that elegant 'Queen Anne's lace', as it is more gracefully known, flourishes along fertile cowpaths?

Another early flowering umbellifer is alexanders, originally introduced from southern Europe. Its large heads of yellow-green flowers can be found in hedgerows and waste places in coastal areas all round the British Isles, although it is commoner in the South. Once widely cultivated as a pot-herb or vegetable, alexanders are a sad ommission from the modern kitchen garden: their fragrant myrrh-like smell and distinctive taste entitle them to high rank amongst culinary plants. They were widely planted in monastic gardens, where they now remain as weeds among the ruins. Since the first century AD it was cultivated like celery and blanched up to minimise its bitterness; the roots were eaten lightly boiled, and the flower heads put into salads. It was known in mediaeval times as 'parsley of Alexandria', and Irish women knew it as an ingredient of 'lenten pottage', a soup of alexanders, watercress and nettles. Parkinson, in his 'Theatricum Botanicum' of 1640, says:

'Our Allisanders are much used to make broth with the upper part of the roote, which is the tenderest part, and the leaves being boiled together, and some eate them either raw with some vinegar, or stew them, and so eate them, and this chiefly in the time of Lent, to helpe digest the crudities and viscous humours [which] are gathered in the stomacke by the muche use of fish at that time.'

In spite of John Evelyn's exhortation to grow it in the kitchen garden and to eat the buds in salad, by the end of the eighteenth century it was superseded by celery.

Its scientific name, *Smyrnium olusatrum*, derives from a Greek word meaning 'myrrh', because of the smell of its juice; *'olus'* is the Latin for vegetable, and *'atrum'* for black, the colour of its ripe fruits. It owes its common name 'alexanders' either to Alexander the Great, who may have discovered it, or to Alexandria, where it may have originated. It may otherwise have come from Macedonia, the birthplace of Alexander the Great.

Alexanders was a recognised medicinal herb for centuries, and the root and seed remained in use until the beginning of the nineteenth century. The seed soaked in wine was believed to promote menstrual bleeding, and the leaves were a remedy for scurvy in days when vitamin C was scarce; the roots are mildly diuretic. 'The leaves bruised and applied to any bleeding wound, stoppeth the blood and dryeth up the fore without any grief,' says William Coles in his sixteenth-century The *Art of Simpling*.

Lords and ladies is quite one of the oddest of British wild flowers. It is almost the sole British representative of a predominantly tropical family of plants containing more than 1,000 species. It is a common sight in shady hedgerows at this time of year, and in order to understand the strange inflorescence it is necessary to peel apart the sheath (spathe) and the base. This will reveal a broad ring of stiff hairs at the base of the central column (spadix). Below this is a ring of male flowers consisting of short-stalked stamens and finally, at the base, there is a zone of female flowers comprising simply ovaries and stigmas. The spathe opens around midday, when, astonishingly, the spadix begins to heat up and give off a scent which attracts a variety of small flies. These fall off the slippery spathe down between the hairs at the base of the spadix and eventually reach the female flowers at the bottom, which they will pollinate if they happen to be carrying pollen from another plant. The female flowers on each plant ripen before the male flowers, which prevents self-pollination. After pollination the male flowers ripen, themselves producing a mass of pollen which gets dusted on to the flies. The stiff hairs then wilt allowing the flies to escape with the chance of visiting another plant. Later in the year the plant again becomes prominent, when the spike of glistening orange-red berries ripens. They should be left severely alone as they are very poisonous.

Adam and Eve, bulls and cows, men and women and lords and ladies all refer to the obvious imagery of *Arum maculatum* (or cuckoopint as it is also known), although Thomas Hardy in *Far from the Madding Crowd* describes it 'like an apoplectic saint in a niche of malachite'. For children the purple-headed spikes were 'lords', the light-coloured ones 'ladies'. By association with its structure Dioscorides claimed that it was aphrodisiac: 'It is uretical too, and stirs up affections to conjunction being drank with wine.' With all its symbolism it is depicted on the Unicorn Tapestries, woven in 1514 as a wedding gift for François I of France.

Also called starchwort and arrowroot, its tuberous roots contain a floury substance which used to be extracted on the Isle of Portland and made into Portland Sago. It also provided a starch used to stiffen Elizabethan and Jacobean ruffs: 'The most pure and white starch is made of the roots of Cuckowpint; but most hurtful to the hands of the Laundresse that hath the handling of it, for it choppeth, blistereth and maketh the hands rough and rugged, and withal smarting,' says Gerard.

John Pechey, the seventeenth-century herbalist, wrote that 'women do use the distilled water of the root to beautify their faces; but the juice of the root, set in the sun, is much better'. A tiny decoction was formerly considered to be a certain remedy for poison and the plague. Its generic name, '*Arum*', comes from a Greek word for poisonous plants, and '*maculatum*' means spotted, describing the leaves which sometimes have purple blotches on them – hence yet another local name, spotted wake-robin.

The brightest splash of colour comes from red campions, which are particularly luxuriant in the hedgebanks of the West country. They are unusual in being one of the few British herbs that have separate male and female plants like the dog's mercury. Red campion has many Robin-names because of its colour, and it is the flower of St Philip and St James, whose feast day is 1 May. Its name was coined by the Romans, who used the plant in garlands to crown their champions in the public games in Rome. Later on it became the 'champion' of sixteenth-century English gardens, and was cultivated to produce the lovely rose campion. Red campion is a flower of the fairy folk, and can cause a death in the family should it be picked, whereas white campion is a thunder flower, and picking it will provoke a storm. Gerard maintained that the campions were an effective remedy for bites: 'The seed drunken in wine is a remedie for them that are stung with a scorpion, as Dioscorides testifieth.'

Also providing colour are the various vetches. The bush vetch, *Vicia sepium*, favours the shadier grassy hedgerows. It has pale purple flowers, rather similar in colour to those of the bitter vetch, *Lathyrus montanus*. However, they can easily be told apart, as the leaves of the bitter vetch have only two to four pairs of leaflets and end in a short point, whereas the leaves of the bush vetch have five to nine pairs of leaflets and end in a long, twining tendril. Bitter vetch prefers well-drained, acid hedgebanks, so is not to be expected in chalk and limestone areas. It is also rare over a large part of the Midlands and is virtually absent from East Anglia.

Stitchwort is a plant to cure a stitch in the side: to the Anglo-Saxons and Celts such a pain was likely to be caused by elf-shot, delivered by the elves to whom this flower belongs. Gerard's comment is: 'They are wont to drinke it in wine with the powder of Acornes, against the paine in the side, stitches, and such-like.' Stitchwort is a shade-loving plant which soon wilts after it is picked: it has weak, brittle stems and lovely white, star-like flowers which have many names: Star of Bethlehem, shirt buttons and adders' meat. Stitchwort grows in grassy habitats where snakes may be found, and in Cornwall children feared that should they pick the flowers adders would bite them. It is a flower protected by the Devil, a thunder-flower which if gathered would provoke thunder and lightning. It was also believed that its medicinal use would ensure the production of male children.

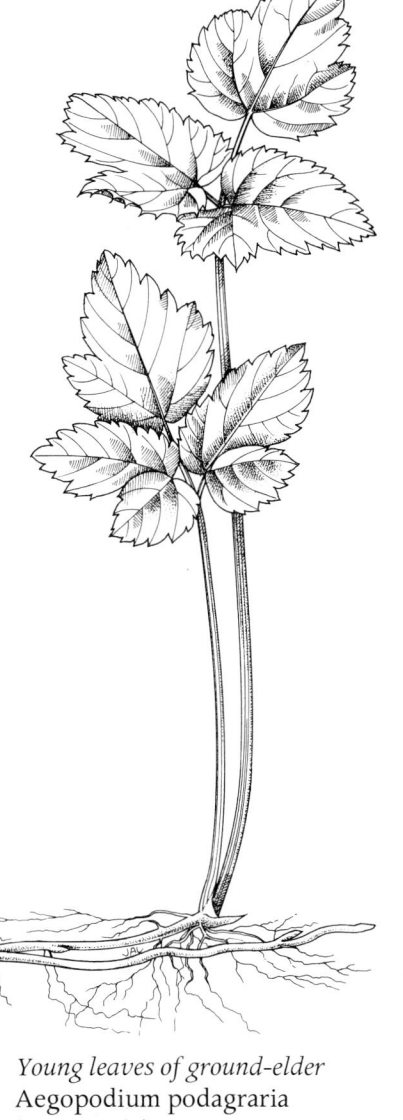

Young leaves of ground-elder
Aegopodium podagraria
(*see page 50*)

May Recipes

Ground Elder Soup
Alexanders Vol-au-Vent with
Creamed Plantain
Blossom Petal Ice Cream
May Blossom Wine Cup
Hawthorn Nibbles
Hop-shoot Fritters

Ground Elder Soup

Serves 2

1 large handful of
    ground elder leaves
50 g/2 oz butter
25 g/1 oz flour
275 ml/$\frac{1}{2}$ pint stock
150 ml/$\frac{1}{4}$ pint milk
cream, salt and pepper

Wash the leaf stems and cook them in melted butter for ten to fifteen minutes, stirring frequently, until tender. Add the flour, and stir in the hot stock gradually until the mixture thickens. Liquidise. Add the milk, heat gently and add seasoning to taste (if indeed it needs any). Finish by pouring the cream on top and serve hot.

Alexanders Vol-au-vent

Boil the young leaf stems of alexanders until they are tender, about ten minutes, and chop them up. Mix with a little béchamel sauce and put into vol-au-vent cases. Cook at 375°F, 190°C or gas mark 5 for twenty minutes.

Creamed Plantain

Pick young plantain leaves and wash them. Cook in salted, boiling water for ten to fifteen minutes until tender, and then drain them. Add to some well-seasoned béchamel, stir well and serve hot.

Blossom Petal Ice Cream

petals of crab-apple blossom,
cherry, pear, etc
150 ml/¼ pint double cream
40 g/1½ oz icing sugar
2 tbs orange flower water
2 egg whites

Infuse the petals in the warmed orange flower water and leave to soak for an hour. Whip the cream with the sifted sugar until it is thick, then add the petals and the orange flower water. Fold in the stiffly beaten egg whites and freeze.

May Blossom Wine Cup

1 bottle white wine
1 bottle red wine
1 orange, sliced
sugar to taste
fresh herbs such as
    thyme, borage, mint, etc
May blossoms, washed

Mix the first five ingredients together, stir well, and scatter the well-washed may blossoms over the top. Cover with a cloth and leave to stand for twenty-four hours. Serve well-chilled, with ice, on May Day morning.

Hawthorn Nibbles

Pick some young hawthorn shoots and leaves, and wash them well. Dry them and put on to pieces of fried bread. Top with a slice of soft cheese, and grill until the cheese is bubbling.

Hop-shoot Fritters

Fritter Batter
110g/4 oz plain flour
a pinch of salt
3 tbs vegetable oil
150 ml/¼ pint warm water
1 egg white

Wash some young hop shoots and dry them thoroughly. Dip into fritter batter (see below) and deep-fry them in very hot oil until golden all over. Drain on kitchen paper and serve immediately, sprinkled with salt.

Sieve the flour and the salt, and stir in the oil. Gradually add the water, and stir it until it becomes smooth and creamy. Leave in a cool place for about two hours and then fold in the stiffly beaten egg white.

June

> *Now summer is in flower and natures hum*
> *Is never silent round her sultry bloom*
> *Insects as small as dust are never done*
> *Wie glittering dance and reeling in the sun*
> *And green wood fly and blossom-haunting bee*
> *Are never weary of their melody*
> *Round field hedge now flowers in full glory twine*
> *Large bindweed bells wild hop and streakd woodbine*
> *That lift athirst their slender throated flowers*
> *Agape for dew falls and for honey showers*

John Clare, 'The Shepherd's Calendar'

1 *Comfrey*
2 *Elder*
3 *Dog Rose*
4 *Field Rose*
5 *Germander Speedwell*
6 *Tufted Vetch*

This month is named after Juno, the wife of Jupiter: maybe then it is the queen of months. The hedgerows are lush with wild roses, elderflowers, honeysuckle and guelder rose, and along the banks countless wild plants are in full flower: clover, vetches, moon daisies and hedge woundwort, exquisite grasses of all kinds, meadowsweet with its heady scent – all creating a bonanza for bees, who are tirelessly collecting pollen and nectar.

It is a month of both Christian and pagan festivals: the feast days of the Trinity and of Corpus Christi fall in June, and the eleventh is St Barnabas' Day, said to be the time for the start of the hay-making:

On St Barnabas'
Put the scythe to the grass.

The fifteenth is the feast day of St Vitus, who is the protector of epileptics and those afflicted with chorea, the disease whose symptoms include the 'dance' named after him.

If St Vitus' day be rainy weather
It will rain for forty days together.

Folklore has another tip for the farmer – if the cuckoo sings after St John's Day, the twenty-fourth, then the harvest will be late.

The major pagan festival of the year falls on St John's Eve, the twenty-third, which is usually called Midsummer Eve – although the summer solstice, the longest day, is on the twenty-first. As well as being one of the Quarter Days in England, when rents are due, it is a highly magical night whose traditional ceremonies probably date back to the times of the Druids. All over the land bonfires were lit at nightfall and kept burning until after midnight, as a symbolic attempt to boost the power of the sun as it started to ebb. As the flames dwindled, people would leap over the fires in the belief that it would bring them good luck and protect them from witchcraft and evil spirits. A gentler ritual aimed at protection from spells and lightning was the gathering of St John's Wort on Midsummer Eve for hanging over the doors and windows of houses and cottages.

June is rich in weather lore because it is an important time for the crops: 'a leak in June brings harvest soon', and 'a dripping June sets all in tune'. Certainly sudden and torrential thunderstorms are typical of the month, and rain on 8 June foretells a wet harvest. But whatever the weather, it is always a busy month for the farmer:

Full summer comes; June brings the longest day.
All country dwellers know the small despair
Of the year's summit; but the yeoman now
Has little time for vain regrets to spare.
There's work enough for him and all his folks.

V. Sackville West, 'The Land'

Documentary history of the Essex and Sussex hedges

It is only comparatively recently that historical geographers have begun to show us just how ancient the landscapes that surround us are. The unravelling of the story of a familiar piece of countryside is an exciting and challenging enterprise. Information from a whole range of different sources all contribute to the solution of the puzzle. Archaeological evidence, aerial photography, place names, documentary sources and soil, plant and animal studies all provide individual pieces from which the jigsaw of the developing scene back through the centuries to the original woodland clearances can be constructed.

Last month our survey of the Essex and Sussex hedgebanks suggested that they were both of considerable antiquity, the bank and ditch of the Sussex hedge being possibly as old as the parish boundary itself. Documentary evidence on the past appearance of the countryside can be found going back to pre-conquest days. Many Anglo-Saxon charters from the ninth century onwards contain records not only of hedgerows but also of individual trees marking boundaries.

The Domesday Survey of 1086 is less helpful over details of individual boundaries, but from the middle of the thirteenth century the amount of documentary evidence increases enormously. A huge amount of fascinating information still remains hidden in estate accounts, local surveys, custumals and court rolls waiting to be interpreted.

Large-scale estate maps began to appear towards the end of the sixteenth century, and many of these show hedges along the field borders and farm boundaries. The numbers of these increased as the enclosure movement gathered momentum during the seventeenth century, and most English parishes were covered by enclosure or tithe maps during the eighteenth or nineteenth centuries. The best place to begin a search for archival evidence is the County Record Office. The County Archivist or Record Officer is invariably only too pleased to assist the genuine researcher and to share his professional experience and enthusiasm. Several counties have their own Record Society which publishes a series of Proceedings in which old documents are reproduced and translated and in which the source references are quoted, such as to the Public Record Office or British Museum Library. For many counties the still incomplete *Victoria County Histories* are an invaluable source of information, especially on such subjects as ecclesiastical history and for Domesday translations.

We enlisted the help of both the Essex and the East Sussex County Archivists in our search for the origins of our two hedges. The tithe map for Pebmarsh was published in 1839, and clearly shows the boundary between the two fields, Rowley and Reedan, on either side of our hedge, although at that time Reedan was divided into two, presumably by a hedge, making a long, narrow field running parallel to our hedge. A check on successive editions of the Ordnance Survey revealed that that hedge was still there in 1954, but had disappeared by 1974. The present owner of Spoon's Hall has in his possession a copy of a map of the farm entitled 'The Manor Farm of Spoon's Hall' and dated 1807, and we were fortunate enough to track down the original in the County Record Office. It has a delightful drawing of an early nineteenth-century surveyor and his assistant in the bottom right-hand corner, complete with level, tripod and levelling staff. There is also an inset drawing of Spoon's Hall which the emblazoned title proclaims to be the property of one Daniel Giles, Esq. However, by far the most interesting feature of the map from our point of view is that Reedan field was not then part of the farm, but labelled as belonging to 'John Stebbing', who appears to have been the local church warden. This means that our hedge was then the farm boundary, and as such it might mark the original limit of the estate.

We then discovered a second and earlier map of 'Spoon Hall Farm and Wood' dated 1771. At that time the estate belonged to Lord Grimston, and our hedge is still clearly the farm boundary. It therefore seems likely that the hedge does mark the original farm boundary, which will be as old as the farm itself. We found no more estate maps, but there are two rental references to Spoon's

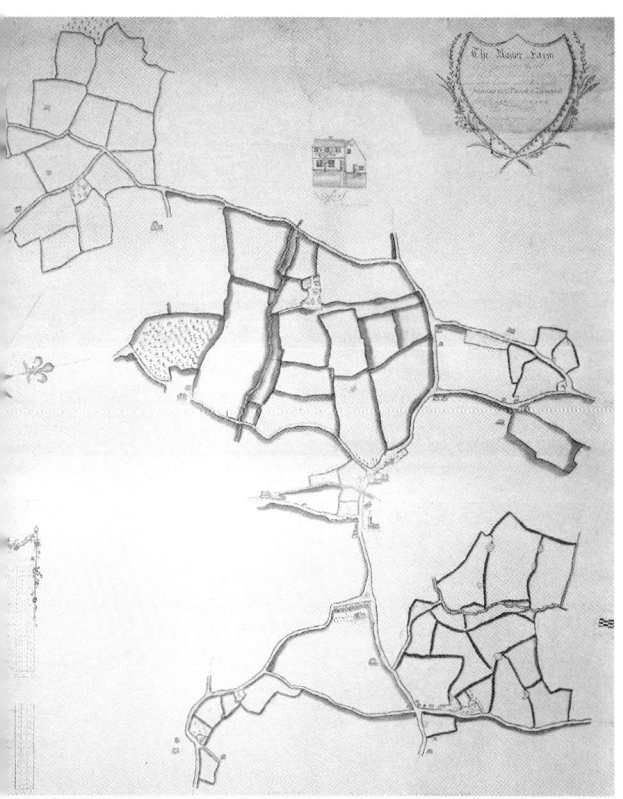

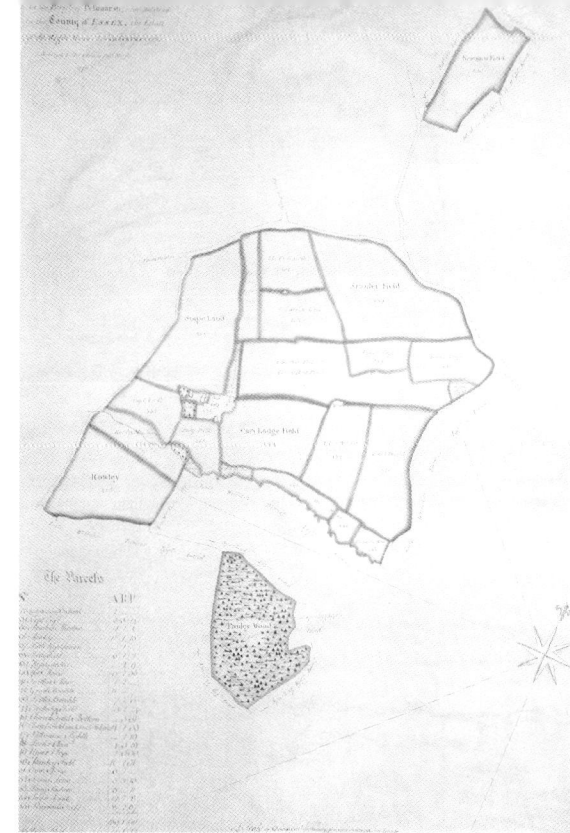

Above
*Map of Spoon's Hall Farm,
Essex, in* 1807
Above right
*Map of Spoon's Hall Farm,
Essex, in* 1771
Below
*Drawing of two surveyors
on the* 1807 *map*

Hall in 1508 and 1541. A solution of the puzzle about the origin of the name of the farm is suggested in a document in the Public Record Office.

A 'Hugh Spon' of Pebmarsh is mentioned in an assize court roll of 1247, so it is a reasonable assumption that either he or one of his forebears is commemorated in 'Spoon's Hall'. Not only does this give us the origin of the name but it also indicates that the farm was already in existence in the mid-thirteenth century. Whether or not the boundary was hedged at that time we have no means of telling from the documentary evidence, and we shall have to use other kinds of information to throw light on that question. Neither were we able to discover when that part of the parish was enclosed, although the straight sides and regular shapes of the fields of the 1771 map do suggest pre-Parliamentary enclosures, some time during the preceding two hundred years.

An interesting insight into one aspect of local agricultural history is provided by the intriguing inclusion of a small 'hop garden' just to the north of the stream at the western end of the Moors on the 1771 map, which had apparently disappeared by 1807. Although hops haven't been grown in the area this century, this part of Essex is known to have been an important hop-growing region in the eighteenth century and indeed Castle Hedingham, four miles away, had its own hop market. What makes the story more fascinating is that when we searched the hedgerows surrounding that piece of land we did discover several patches of hops growing wild, scrambling over the bushes, that have clearly persisted as relics of cultivation for about two hundred years.

In contrast to the Essex hedge, our survey of the physical structure of the Sussex hedge together with its identification with the parish boundary had already led us to suspect that it was of considerable antiquity. The name of the farm, Pinehurst, is almost certainly of Saxon origin. 'Hurst' is a common place-name suffix in the Central Weald of Kent and Sussex and is derived from the Saxon *hyrst*, 'a wooded hill'. The name thus appears self-explanatory. The only problem is that it is generally accepted by historical ecologists that pine died out in lowland Britain long before the Saxon period and did not reappear as a feature in the landscape until it spread from cultivation in the seventeenth century. Therefore, at first sight Pinehurst is rather an enigma but Catherine Pullein in her book on the history of Rotherfield records two late sixteenth-century documents in which the farm is referred to as 'Pyndhurst'. Now the Old English for 'pine' was *pin*, but on the other hand in the sixteenth century a 'pynder' was an officer of the manor who had the responsibility of impounding stray animals. Furthermore, 'pynder' is itself derived from the Saxon *pyndan*, 'to shut up or enclose'. Farms in this part of the world are frequently older than the villages and Pinehurst thus appears to refer to a wooded hill on which animals were enclosed; a name that could have its origins as far back as the sixth or seventh century.

Section of the map of the Manor of Rotherfield showing part of the southern boundary incorporating the Sussex hedge

At this stage we discovered that the East Sussex County Record Office possessed a real treasure: a beautiful hand-coloured map of the Manor of Rotherfield executed in 1664 as a copy of a much earlier map dated 1597. The map is a landscape historian's dream. Each individual field is marked and colour-coded to correspond with its origins and tenure – assart, demesne and ferlings. But what was exciting for us was the realisation that not only was our hedge the parish boundary between Rotherfield and Hadlow Down, but as Hadlow Down was situated in the neighbouring manor of South Malling, it was an ancient manorial boundary as well. The history of the manor of Malling is well documented. It was originally granted to the Archbishop of Canterbury by King Bealdred of Kent in about 823, and the grant was subsequently confirmed by King Ecgberht and his son Aethelwulf at the Council of Kingston in 838. Our hedge thus occupies part of the boundary of an ecclesiastical Saxon manor.

Of course, this does not necessarily mean that that part of the manor was permanently settled as early as the ninth century; indeed, it is rather unlikely that it was. However, research by Dr Peter Brandon has led him to conclude that the area bounded by our hedge was settled sometime in the period between the Domesday Survey of 1086 and 1220. Whether or not the limits of the Archbishop's manor were delineated by a physical boundary at the time of clearance or earlier is a matter of speculation. That a physical manorial boundary did exist at an earlier date is suggested by a charter of 1018, recording a grant of land at the northern end of the manor by King Canute to Archbishop Aelfstan, the north-western limit of which ran down 'the broad stream by the Arch-bishop's boundary'.

Taken together, our documentary researches suggest that both of our hedges are probably of considerable antiquity. The Essex hedge forms part of the boundary of a farm that was in all likelihood in existence in the mid-thirteenth century, whilst the Sussex hedge marks the limits of a ninth-century ecclesiastical manor. Further evidence, of a different kind, will have to be used to help decide whether the hedges are as old as the boundaries that they mark.

Grasses

By midsummer the hedge-bottom is overgrown by grasses – and very attractive many of them are. They are by far the commonest of our wild flowers, although to some the term 'flower' may not seem particularly apposite. Many are in full bloom during June, as anyone who suffers from hay fever will know to their great discomfort. Altogether there are between 150 and 160 different species in Britain, and of these about two dozen can be found growing along the hedgerow. Some of these will have spread from the neighbouring meadows; others are more characteristic of shady woodlands. Some species are commoner on lime-rich soils, others on acid. A few, such as the oat-grass, *Arrhenatherum elatius*, may actually now be most at home in the hedgerow.

Most grasses are not at all difficult to name, and a key to help with the identification of the commoner hedgerow species is on page 64. The flowers are very much reduced, lacking the showy petals and sepals of more conventional flowers, so that each simply consists of three stamens and a central ovary crowned by two feathery stigmas. Each of these little flowers or florets is enclosed within a pair of pale-coloured scales, the lemma and the palea. The florets in turn are grouped together in spikelets, and it is the arrangement of the spikelets in the inflorescence that gives each grass its own particular appearance. The illustration on page 64 shows a typical spikelet together with a number of representative inflorescences. Most grasses can be identified quite confidently, even when they are not in flower, by examining the leaves and the shoots and sometimes the roots. The leaves are composed of two parts: the leaf-blade, which is the obvious bit, and a basal sheath, which encircles the stem. Where the two join there is a small scale, called the 'ligule', which normally lies tight against the stem. The shape and size of the ligule is a particular help in identification, and it can quite easily be seen by gently pulling the base of the leaf blade away from the stem. Other points to notice are whether any parts of the plant are hairy, the arrangement of the veins as they appear on each surface of the leaf and the shape of the leaf tip – for example, whether it is flat or boat-shaped. Grasses make a splendid subject for study, and the hedgerow is an ideal place to start in getting to know the group of plants on which almost the whole of mankind depends for its food. Wheat, barley, oats, rye, rice and maize have all been developed from wild grasses to produce the staple crops on which we depend. Grasses dry and press well and also make attractive subjects for the dry flower arrangement. What is more, a collection of hedgerow grasses is unlikely to cause any concern from the conservation point of view, providing that it is remembered that plants should only be picked and not uprooted.

One of the first grasses to come into flower is the meadow fox-tail, *Alopecurus pratensis*, the yellow or purple anthers covering the long, silky inflorescence spikes as early as April. As its name suggests, it is really a plant of the pasture, but it is often to be seen in grassy hedgebanks bordering fields or road verges. The same is true of the common rye-grass, *Lolium perenne*, and the crested dog's-tail, *Cynosurus cristatus*, both of which also have spike-like inflorescences. Several of the larger, more robust grasses commonly occur along the roadside hedge such as oat-grass, *Arrhenatherum elatius*, barren brome, *Bromus sterilis*, and the cock's-foot, *Dactylis glomerata*, the inflorescence of which is supposed to resemble the foot of a fowl – but that requires some effort of the imagination! Shadier hedgerows, not surprisingly, acquire more characteristic woodland species like giant fescue, *Festuca gigantea*, hairy brome, *Bromus ramosus*, and wood false-brome, *Brachypodium sylvaticum*, whilst a leafy bank may be carpeted by the attractive wood melick, *Melica uniflora*.

Cuckoo-spit

Among the most familiar of midsummer sights are the little patches of foam, known as cuckoo-spit, that appear on a whole host of hedgerow plants. If the foam is smeared flat, a tiny yellowish creature will be found lurking in the middle. It is the larva of a very abundant insect, the common frog-hopper, *Philaenus spumarius*.

The adult frog-hopper is about 5 mm long, and mottled brown in colour, the actual patterning and intensity of the colour varying from place to place. In any event, it is well camouflaged amongst the general herbage. It belongs to the order of insects that also includes the aphids and scale insects, as well as the tropical cicadas, the Homoptera. They feed by sucking plant juices through their sharp, piercing mouth-parts. As their name implies, their enlarged hind legs enable them to hop, although whether or not they really bear any resemblance to miniature frogs must be left to the individual imagination!

The cuckoo-spit is thought to protect the larva both from potential predators and also from desiccation while it sits in the middle of its bubble-bath feeding on the plant sap. It is certainly true that, if removed from the foam, they quickly dry up and die, but whether it forms a very efficient cover from predators is another matter. Some species of solitary wasp certainly collect large numbers of the larvae by fishing them out of the enveloping spume and carrying them off to their nest to feed their own larvae.

The way in which the insect actually produces the foam was for a long time a matter of speculation. It appears that the side walls of the frog-hopper's abdomen are extended beneath the body to meet along the middle of its underside, so enclosing a cavity into which

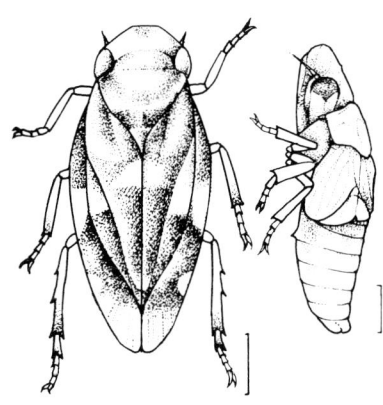

Common frog-hopper Philaenus
spumarius
Left: adult
Right: nymph, side view

its breathing pores open. This cavity is closed at the front end, but allows air in or out by way of a valve at the rear end. The froth is produced by a fluid issuing from the anus forming a film across the valve. Air is then expelled through the valve, blowing the fluid into bubbles. An extraordinary property of cuckoo-spit is that it maintains its coherence even in wet weather. Indeed, it would be of little value to the larva if it were washed away every time that it rained! The trick is performed by glands at the rear end of the insect's abdomen which produce a sticky substance that is mixed with the froth giving it its viscid consistency.

There are several other widespread species of frog-hopper in Britain, the most striking of which is undoubtedly the scarlet and black *Cercopis vulnerata*, which openly flaunts itself on its food plants. These include several of the common plants of damper hedgerows. The suggestion is that, like many other gaudily attired insects such as the burnet-moths, it is distasteful, so that its distinctive self-advertisement has evolved as a warning to potential predators. Unlike *Philaenus*, the larvae live gregariously underground in a mass of solidified froth. The adult frog-hoppers of both species are common throughout mid and late summer.

June flowers

The wild roses are undoubtedly the most conspicuous and the most beautiful of the midsummer hedgerow flowers. There are more than a dozen different species in Britain, almost all of which occur in hedges, but by far the commonest are the dog-rose, *Rosa canina*, and the field rose, *Rosa arvensis*.

Our common dog-rose occurs throughout the British Isles and extends right through Europe and North Africa to south-west Africa. It is an extremely variable plant, and sometime in history it was crossed with the damask rose, *Rosa damascena*, to produce the White Rose of York. The flowers are white or, more usually, pink, and the stigmas form a short, round dome-shaped group in the middle of the flower. The flowers of the field rose are nearly always white, but a more certain way of distinguishing it from the dog-rose is to look at the styles, which are all joined together to form a column in the middle of the flower that is as long as the stamens. Furthermore, the field rose is a much more scrambling plant than the dog-rose and, indeed, belongs to the group of roses that has given rise to most of our modern ramblers. The beautiful fragrance is given off by the stamens, not the petals.

In chalky districts in southern England and on lime-rich soils scattered about the rest of the country you may come across the sweet-briar. The flowers are a deep pink, and the leaves when torn and crushed smell strongly of apples.

In northern Britain, on the other hand, the downy rose becomes more abundant. Also usually deep pink, the leaves are covered with short, downy hairs which gives them a rather dull appearance.

The wild rose flourished in ancient Athens, and in the *Iliad* it features as the flower of Aphrodite, the goddess of love, who

Above left
Rye-grass Lolium perenne
Centre left
Crested dog's-tail Cynosurus
cristatus
Below left
Cock's-foot Dactylis glomerata

61

cured Hector's wounds with the oil of roses. The rose descended from Aphrodite to Eros and then became the sacred Rosa Mystica of the Virgin Mary. The island of Rhodes gets its name from the roses that were cultivated there.

With its contrasts of sharp thorns and delicate, scented flowers, the wild rose has become the symbol of that true love which survives happiness and adversity alike, for it symbolises both pleasure and pain. Shakespeare has an epithet along those lines: 'Roses have thorns and silver fountains mud.' To dream of being pricked by briars (another name for the dog-rose) indicates that the dreamer is in love, and striving to gather his rose through the thorny bush. The beauty of this delicate wild plant has been admired in literature and poetry for centuries and is a much-loved feature of our hedgerows. Rupert Brooke's couplet is almost legendary:

Unkempt about those hedges blows
An English unofficial rose.

Creamy clusters of elderflowers not only look beautiful, they also have a heady scent, and give a muscatel flavour to anything they are mixed with. It has been said that the smell is narcotic and that it is unwise to plant an elder tree near a bedroom window. Likewise it is unwholesome for cattle to rest in its shade. It can cause dizziness, and there is an old saying that 'he who sleeps under an Elder-tree will never wake'. Farm carts used to be drawn by horses whose bridles had sprays of elderflowers attached to them in order to deter the flies, and dried elderflowers make a good insect repellent. French fruit growers have been known to store pear crops in elderflowers for a period of months so that their muscatel flavour would permeate the fruit. The blossoming of the elder tree has a message for sheep farmers:

If verdant elder spreads
Her silver flowers . . .
Gay shearing time approaches.

There is a northern European belief that if the flowers were put in ale, and a man and a woman drank it together, they would be married to each other within the year.

Elderflower water, which is an astringent, still has a place in the British Pharmacopoeia as eye and skin lotions. It can be used as an aftershave or for soothing sunburn, and has many cosmetic uses in beauty preparations as well.

Short for 'speed you well', speedwell was believed to give protection on a journey and to bring good luck and good health: in Ireland travellers would wear speedwell to protect them from accidents. Alternatively, its name could be a synonym for 'fare-well', since the flowers fall off as soon as they are picked. In the medicinal sense it means 'prosper well, get well', and had a reputation for healing: 'the water of Veronica distilled with wine, and re-distilled so often until the liquor wax of reddish colour,

Bumblebee on comfrey

prevaileth against the old cough, the dryness of the lungs, and all ulcers and inflammations of the same', says Gerard.

The common hedgebank speedwell is the germander speedwell which Ann Pratt, writing about flowers in the last century, describes as 'brilliant blue blossoms lying like gems among the bright May grasses', and it has other names like Billy bright-eye, cats' eyes, eyebright and wish-me-well. It is a species of *Veronica*, named after the saint who wiped the sweat from Christ's brow as he was carrying the cross: the markings on the petals are a representation of the traces left on her handkerchief. '*Vera icon*' means true image, and so speedwell came to represent fidelity. '*Chamaedrys*', its specific name, comes from two Greek words, one meaning 'on the ground' or 'dwarf', the other 'oak', from the resemblance of the shape of its leaves.

Vetches continue to be prominent in the hedgerows in June. The dense-flowered inflorescence of the tufted vetch, *Vicia cracca*, with its blue-purple flowers scrambles up through the hedge, whereas the purple-flowered common vetch, *V. sativa*, prefers the grassy Hedgebanks. The vetches have a reputation for indestructibility:

A vetch will grow through
The bottom of an old shoe.

It is excellent at strangling crops, and Gerard remarked of it, 'The herbe is better knowne than desired'. Common vetch was introduced from Asia by farmers for their cattle, and the seed is much sought after by pigeons.

The commonest comfrey of roadsides and hedgerows is the blue or Russian comfrey, *Symphytum × uplandicum*, with flowers that start off pink and end up blue. It prefers fairly dry conditions, being replaced in damper places by the similar-looking common comfrey, *S. officinale*, which differs in that the bases of the leaves continue down the stem as broad wings on either side.

Comfrey was one of the chief plants used in early medicine for healing wounds and binding fractures; its generic name *Symphytum* is from the Greek meaning 'grow together', and it was grown in old gardens as a necessity of life; indeed, the Pilgrim Fathers took white comfrey with them to the New World. It was used externally in the form of compresses: the leaves were used to heal cuts and wounds, and to ease bruises and gout; and the roots as a poultice for setting bones, hence its local names knitbone and boneset. Tea made from the leaves is good for chest complaints and juice from the root, mixed with sugar and liquorice, can cure a cough. It is a valuable food plant and has been eaten by man for centuries. It has a higher protein content than spinach and contains all the latter's goodness.

Cow parsley is now over, but in many places, especially on soils containing some lime, a similar plant but with rough, purple-spotted stems is now in flower. This is rough chervil. Never as abundant as cow parsley, it nevertheless occurs throughout most of the British Isles, but becomes rare in Scotland and Ireland.

A key to the commoner grasses of the hedgerow

Starting at number 1, decide which of each pair of descriptions fits the plant to be named. Then go to the number indicated on the right hand side and continue until the plant is identified.

Use the illustrations in conjunction with the key. Please note that it will not be possible to name the plant from the drawings alone; read together with the account on page 59.

1	Inflorescence unbranched, or stalks of spikelets very short so that inflorescence appears as a tight cylinder or spike-like	**2**
	Inflorescence obviously much branched when mature	**10**
2	Individual spikelets attached direct to inflorescence stem	**3**
	Spikelets attached in groups to main stem, forming a dense cylindrical inflorescence	**5**
3	Spikelets attached broadside-on to main stem	**4**
	Spikelets attached edgeside-on to main stem. Leaves shining green and with a distinct midrib beneath, dull with numerous parallel veins on upper surface	**Rye-grass** (diagram 5) *Lolium perenne*
4	Leaves pale yellow-green and hairy. Spikelets with long awns. Shady hedgebanks	**Wood false-brome** *Brachypodium sylvaticum*
	Leaves scarcely hairy. Spikelets without awns	**Couch grass** *Agropyron repens*

1 *Meadow fox-tail*
2 *Cock's-foot*
3 *False oat-grass*
4 *Single spikelet*

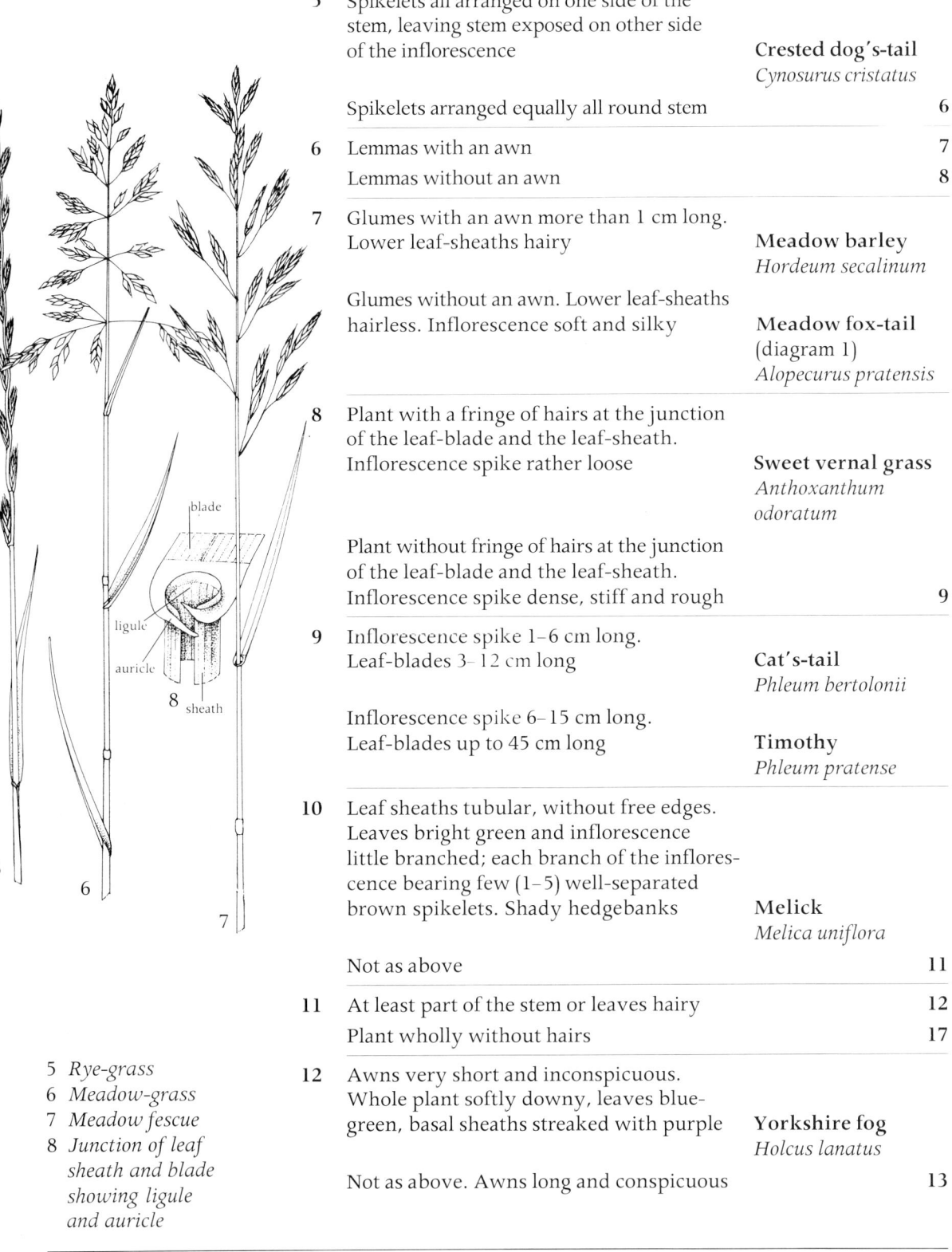

| 5 | Spikelets all arranged on one side of the stem, leaving stem exposed on other side of the inflorescence | **Crested dog's-tail** *Cynosurus cristatus* |
| | Spikelets arranged equally all round stem | 6 |

| 6 | Lemmas with an awn | 7 |
| | Lemmas without an awn | 8 |

| 7 | Glumes with an awn more than 1 cm long. Lower leaf-sheaths hairy | **Meadow barley** *Hordeum secalinum* |
| | Glumes without an awn. Lower leaf-sheaths hairless. Inflorescence soft and silky | **Meadow fox-tail** (diagram 1) *Alopecurus pratensis* |

| 8 | Plant with a fringe of hairs at the junction of the leaf-blade and the leaf-sheath. Inflorescence spike rather loose | **Sweet vernal grass** *Anthoxanthum odoratum* |
| | Plant without fringe of hairs at the junction of the leaf-blade and the leaf-sheath. Inflorescence spike dense, stiff and rough | 9 |

| 9 | Inflorescence spike 1–6 cm long. Leaf-blades 3–12 cm long | **Cat's-tail** *Phleum bertolonii* |
| | Inflorescence spike 6–15 cm long. Leaf-blades up to 45 cm long | **Timothy** *Phleum pratense* |

| 10 | Leaf sheaths tubular, without free edges. Leaves bright green and inflorescence little branched; each branch of the inflorescence bearing few (1–5) well-separated brown spikelets. Shady hedgebanks | **Melick** *Melica uniflora* |
| | Not as above | 11 |

| 11 | At least part of the stem or leaves hairy | 12 |
| | Plant wholly without hairs | 17 |

| 12 | Awns very short and inconspicuous. Whole plant softly downy, leaves blue-green, basal sheaths streaked with purple | **Yorkshire fog** *Holcus lanatus* |
| | Not as above. Awns long and conspicuous | 13 |

blade
ligule
auricle
8 sheath

5 *Rye-grass*
6 *Meadow-grass*
7 *Meadow fescue*
8 *Junction of leaf sheath and blade showing ligule and auricle*

| 13 | Awns bent. Spikelet with 2–4 florets | 14 |
| | Awns straight. Spikelets with more than 4 florets. Leaf-sheaths densely hairy | 15 |

| 14 | Awns more than 1 cm long | **False oat-grass** *Arrhenatherum elatius* | |
| | Awns less than 1 cm long. Inflorescence a silky yellow-green | **Yellow oat-grass** *Trisetum flavescens* | |

| 15 | Leaves more than 7 mm wide, auricled and dark green. Robust grass of shady hedge-banks on moist soils. | **Hairy brome** *Bromus ramosus* | |
| | Leaves less than 7 mm wide, not auricled | | 16 |

| 16 | Spikelets on long drooping branches. Robust grass of hedgerows and waste places. | **Barren brome** *Bromus sterilis* | |
| | Spikelets crowded and erect | **Soft brome** *Bromus mollis* | |

| 17 | Leaves keeled and with a boat-shaped tip | 18 |
| | Leaves with a flat tip | 21 |

| 18 | Shoot distinctly flattened at the base. Inflorescence of characteristic shape (See page 64) | **Cock's-foot** (diagram 2) *Dactylis glomerata* | |
| | Shoot not distinctly flattened at the base | | 19 |

| 19 | Ligule shorter than broad | **Meadow grass** (diagram 6) *Poa pratensis* | |
| | Ligule longer than broad | | 20 |

| 20 | Plant less than 25 cm tall | **Annual meadow grass** *Poa annua* | |
| | Plant more than 25 cm tall | **Rough meadow grass** (diagram 6) *Poa trivialis* | |

| 21 | Basal leaves dark green, long and bristle-like | **Red fescue** *Festuca rubra* | |
| | Leaves flat, not bristle-like | | 22 |

| 22 | Leaves with a single prominent midrib on the upper surface | 23 |
| | Leaves with numerous very fine ribs on the upper surface (look carefully) | 24 |

| 23 | Leaves with distinct auricles. Spikelets awned. Large robust grass of shady wooded hedgebanks. | **Giant fescue** *Festuca gigantea* |
| | Leaves without auricles. Spikelets without awns | **Meadow fescue** (diagram 7) *Festuca pratensis* |

| 24 | Spikelets awned | **Brown bent** *Agrostis canina* |
| | Spikelets without awns | 25 |

| 25 | Ligules of non-flowering shoots shorter than wide. | **Common bent** *Agrostis tenuis* |
| | Ligule of non-flowering shoots longer than wide | 26 |

| 26 | Plant with creeping runners. Inflorescence remains more or less closed up when mature | **Creeping bent** *Agrostis stolonifera* |
| | Plant without runners. Inflorescence spreading when mature | **Black bent** *Agrostis gigantea* |

June Recipes

Potato and Mallow Soup
Comfrey Layers
June Salad
Elderflower and Strawberry Dessert
Elderflower and Orange Cordial
Herb Dip

Potato and Mallow Soup

Serves 2

2 handfuls young
    mallow leaves
25 g/1 oz butter
275 ml/$\frac{1}{2}$ pint stock
225 g/$\frac{1}{2}$ lb mashed potato
salt and pepper
cream to garnish

Shred the mallow leaves finely and cook in the melted butter for seven to ten minutes until softened. Blend the stock with the potato until smooth, then stir in the mallow and butter mixture. Season to taste, heat through and finish with cream. Try freezing it to keep for cold winter nights.

Comfrey Layers

Serves 4

450 g/1 lb comfrey leaves
175 g/6 oz rice
225 g/8 oz cheese
butter
milk
salt and black pepper

Pick the comfrey leaves when they are young and still a fresh green colour. (The larger, darker leaves will be tough and a little bitter.) Wash them well, and cook them in their own water with salt until they are tender and cooked down like spinach, about ten minutes. Cook the rice in boiling, salted water until it is tender. Make layers of the comfrey, rice and cheese, seasoning them well, and finish with a layer of the rice. Dot with butter, and pour milk over the top until the dish is about three-quarters full. Season generously with black pepper and bake at 375°F, 190°C or gas mark 5 for twenty-five to thirty minutes.

June Salad

Pick a bunch of chickweed and wash it well. Cut off the straggly roots. Mix with some bean sprouts and dress with a mixture of soy sauce, garlic, olive oil, salt and pepper.

Elderflower and Strawberry Dessert

5–6 elderflowers
275 ml/$\frac{1}{2}$ pint milk
2 eggs
40 g/1$\frac{1}{2}$ oz sugar
10 g/$\frac{1}{2}$ oz flour, sifted
1 tbs elderflower and
    orange cordial (see below)
knob of butter
25 g/1 oz castor sugar
225 g/$\frac{1}{2}$ lb strawberries

Infuse the elderflowers in the milk over the lowest possible heat for half an hour. Leave to cool. Separate the eggs and beat the yolks with the sugar, and then stir in the flour.

Strain the milk and pour it on to the yolks, beating until the mixture is smooth. Cook over a low heat, stirring until it thickens. Flavour with elderflower cordial and add the butter.

Whisk the egg whites with the castor sugar until very stiff, and fold into the hot custard. Line the bottom of individual glass dishes with halved strawberries, reserving a few for decoration. Pour the custard over them and leave to set for an hour or two. Just before serving, decorate with the reserved strawberries.

Elderflower and Orange Cordial

(See notes on wine-making, page 104)

25 heads of elderflower, washed
1 kg 350 g/3 lb sugar
1 litre 700 ml/3 pints water
50 g/2 oz tartaric acid
1 lemon and 4 oranges, sliced

Put everything into a large pan and stir from time to time over a period of twenty-four hours. Strain and bottle. Dilute to taste.

Herb Dip

To $\frac{1}{2}$ a 150 ml/5 fl oz carton of sour cream add:
8 chopped wild chives
8 stalks of salad burnet
2 top shoots of wild mint
1 small young tansy leaf.
Add salt and pepper to taste, and serve with crisps.

Tansy Tanacetum vulgare

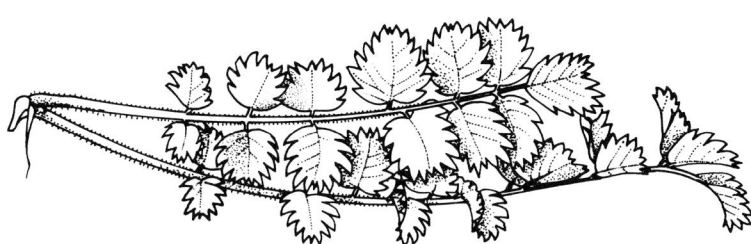

Salad burnet Sanguisorba minor (= Poterium sanguisorba)

July

This is the month of high summer: the trees and hedges are now heavy with green foliage, and the countryside seems somehow to be holding its breath. There is a sense of fulfilment following a time of expectancy. Apart from splashes of bright red poppies along the field borders, much of the colour of the wild flowers is vanishing. White bryony climbs high into the hedges, which are now almost silent apart from the occasional song of the cock yellowhammer. Sadly, this is often a time when indiscriminate early mowing of the verges takes place, interrupting the natural cycle of plant life and causing many flowers, particularly the biennials, to disappear from their habitat before they have time to seed.

1 *Stinging Nettle*
2 *White Dead-nettle*
3 *Cleavers*
4 *Foxglove*
5 *Upright Hedge-parsley*
6 *Broad-leaved Dock*

July is the month for haymaking, but is also a time when we watch the sky for rain and thunderstorms, for although on average it is the warmest month of the year, it can also be a very wet one.

If it rains on St Swithin's Day, 15 July, then it will rain for the next forty days, according to country lore:

St Swithin's Day, if thou dost rain,
For forty days it will remain;
St Swithin's Day, if thou be fair,
For forty days 'twill rain nae mair.

St Swithin was a bishop of Winchester in the ninth century, and the legend goes that until the cathedral was consecrated a hundred years after his death, he was buried outside the church and resented the rain falling on his grave. His time of neglect, however, came to an end when his bones were moved to a shrine inside the newly consecrated cathedral, so we now have to put up with only forty days of his anger!

When the sun does shine, however, it can be very hot, and on a walk along the hedgerows you may be joined by a plant that insists coming with you – goosegrass. With its hairy stems and leaves clinging to your clothing and to the fur of animals, it is making a determined attempt to disperse its seeds.

So-called because geese love eating it, goosegrass used to be fed to goslings in the old days to fatten them up and build up immunity to disease; horses and cows eat it too. In ancient Greece the hairy, sticky leaves were used by shepherdesses to strain hairs out of the milk. Local names for goosegrass include 'cleavers', 'sticky Billy', 'kiss-me-quick' and 'gosling scrotch'.

Its roots produce a red dye, and the roasted seeds have been used as a coffee substitute. Apparently goosegrass was an ingredient of sixteenth-century slimming regimes, presumably as aversion therapy: according to the herbalist John Gerard, 'women do usually make pottage of Cleavers with a little mutton and oatmeale, to cause lankenesse, and to keep them from fatness'.

Pollination

There is a corner on the warm south side of the Sussex hedge which at this time of the year is a mass of the beautiful, pale mauve-pink flowers of the musk mallow, each looking exactly like smaller versions of their exotic Asiatic relative the hibiscus. Contemplation of this blaze of colour inevitably leads one to wonder what purpose is served by the enormous range of extraordinary structures that we call flowers and that have so delighted our poets and artists, gardeners and ramblers.

The truth is that plants have evolved flowers – or, more precisely, the large colourful parts of the flowers, such as the petals – solely for the purpose of attracting a pollinating animal to carry pollen from the flower of one plant to that of another. The animal is usually an insect, but in some parts of the world both birds and bats can act as pollinators. In order to fulfil this function, a flower needs to attract the insect in the first place, and this is achieved by both scent and colour. Secondly, it must provide a suitable landing platform for the insect, and thirdly some inducement to the pollinator to stay and move about in the flower long enough to ensure that the pollen is transferred from the stamens on to the body of the insect. This inducement is usually an offer of refreshment to the guest in the form of nectar, which is simply a syrup comprising a solution of various sugars. In the simpler kind of arrangement as seen in the musk mallow, the petals of the individual flowers provide the landing platform as well as acting as an

73

attraction through both their colour and their scent. The nectar is secreted in special structures, the nectaries, right at the base of the petals, so that to reach it the insects must delve deep into the flower and in so doing will rub against the stamens, their bodies becoming covered with the powdery pollen in the process.

The whole purpose of insect pollination is to secure cross-pollination rather than self-pollination, and many plants have evolved elaborate mechanisms to reduce the chance of self-pollination. In the mallow the stamens ripen before the stigmas of the same flower become receptive, so that by the time the stigma is ripe most of the pollen of the flower will already have been dispersed and it will be pollinated by a visiting insect carrying pollen from a different plant. One of the commonest hedgerow flowers with an arrangement similar to the mallow is the bramble, which is frequented by a whole range of different insects, but unlike the mallow it has no mechanism to reduce the chance of self-pollination.

A further development can be seen in the members of the Umbelliferae, many of which are amongst our most conspicuous and characteristic hedgerow herbs, such as cow parsley, hedge parsley and hogweed. Here the individual flowers are small, but have become massed into large flat heads, the umbels, providing one large 'landing platform' – much bigger than could be achieved by a single flower on a plant of the same size. The flowers of each umbel develop synchronously, so that all the stamens develop and wither before the stigmas mature and become receptive. The alighting area is further increased as the outer petals of the outer flowers of each umbel are enlarged to form an additional flange around the edge of the umbel. An enormous variety of different insects visit these 'floral tables'. Those that we observed on the hogweed at the Sussex hedge included honey bees, bumble-bees, hover flies, the attractive orange and black soldier beetles, small capsid bugs and a strikingly marked yellow and black beetle, *Strangalia maculata*, which is a lovely example of a wasp mimic. In addition, such a host of potential prey insects inevitably attracts predators such as wasps and damsel bugs.

One of the most striking of the seasonal features of the Sussex hedge are the stately inflorescences of the foxgloves that grow all along the north side of the hedgebank. Foxgloves are very choosy about their soil requirements, preferring acid, well-drained soils so that the central Wealden sandstones suit them very nicely. They also have a preference for partial shade: hence their restriction to the shaded north side of the hedge. Unlike the flowers of the mallow, bramble and hogweed, those of the foxglove have evolved to exploit one kind of insect pollinator in particular – the bumble-bee. Each finger-shaped flower is of just a size comfortably to accommodate the body of the bumble-bee, which has to crawl right up the tube in order to reach the nectar at the base of the flower. In so doing, its back will brush against the anthers of the four stamens, which are situated right beneath the upper part of the flower tube. As in the hogweed, the stamens ripen before the stigmas, so ensuring cross-pollination.

74

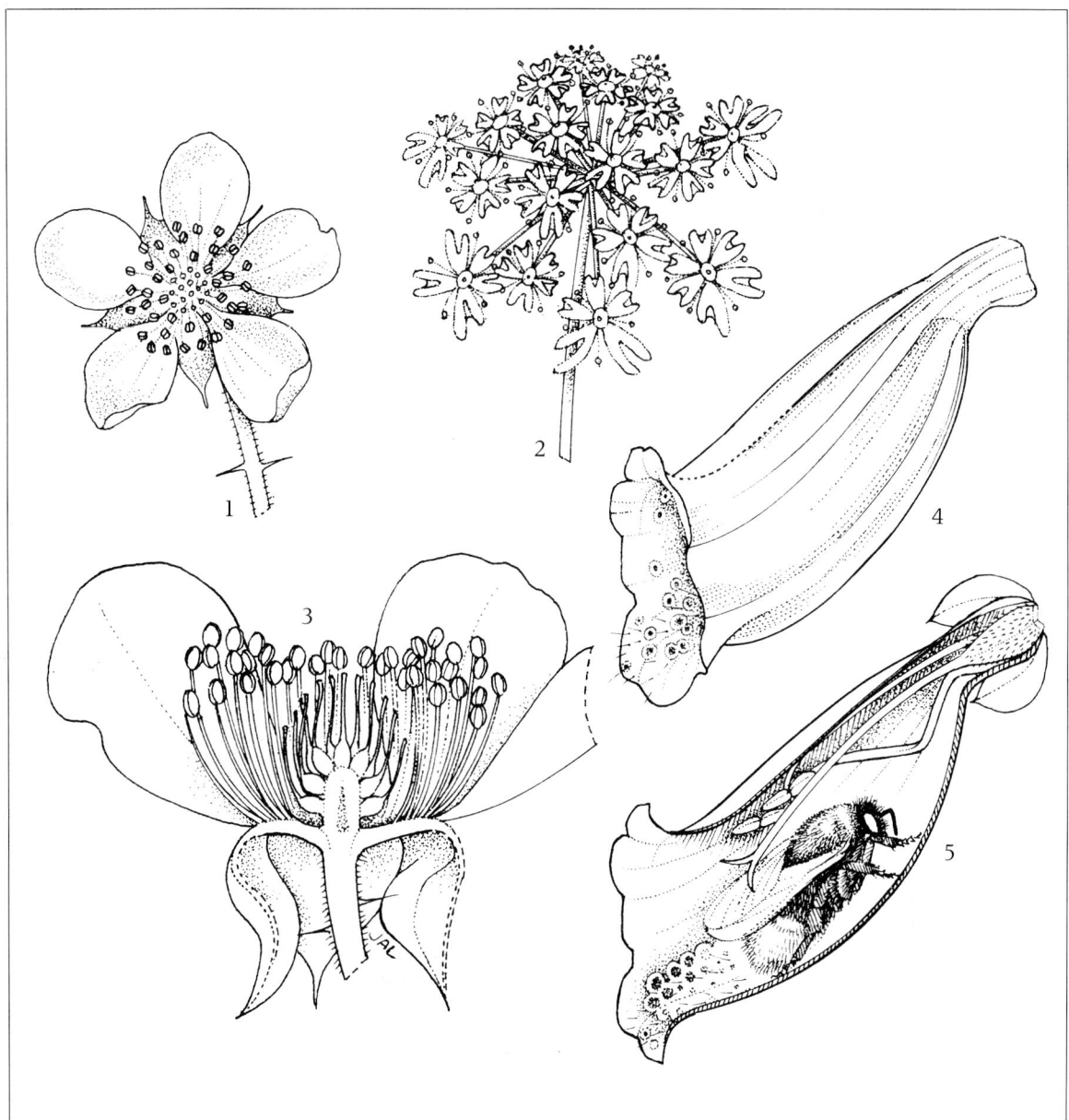

1 *Bramble flower*
2 *Cross-section of bramble flower – the nectaries are situated at the base of the petals*
3 *Hogweed showing the enlarged outer petals of the outer flowers of umbel*
4 *Corolla of the foxglove*
5 *Longitudinal of same with stigma of flower making contact with back of pollinating bumblebee*

Other familiar hedgerow bumble-bee flowers are the white dead-nettle and the hedge woundwort. If the flowers are examined carefully they will be found to consist of two lips, the lower providing the landing platform and the upper forming a hood protecting the style and the four stamens. The lower part of the flower consists of a narrow tube about a centimetre long, leading to the ovary and nectaries. The bumble-bee neatly fills the space between the lip and the hood, and its tongue is exactly the right length to reach the nectar at the base of the tube. At the same time, the bee receives a dusting of pollen on its back, but as the stigmas and anthers ripen at the same time, unlike the foxglove, there does not appear to be any effective barrier against self-pollination.

By far the most frequent and characteristic insect visitors to flowers are, of course, bees, the most familiar of which are the honey bee and the various species of bumble-bee. Bees are unique among the insects in that not only do the adults feed on nectar but the larvae also feed on both nectar (after it has been converted into honey in the bees' crop), and pollen, which is collected by the workers and carried to the nests. Pollen may get deposited almost anywhere on the bee's body, and it then has the problem of transferring it into the pollen baskets situated on the outside of the hind legs, where it can be seen as a yellow, egg-shaped mass. The pollen is brushed off the head and the front end of the thorax by the antennae and front legs, from the rear end of the thorax by the middle legs and from the abdomen by the hind legs. It is then worked into the bristles on the inner side of the middle leg, where it is transformed into a sticky mass by mixing it with regurgitated honey. The pollen mass is scraped into the bristles on the ends of the hind legs. Then follows the most remarkable part of the whole operation. One of the joints of the hind legs is hollowed out to form a 'pollen-press'. The pollen mass from one leg is transferred into the pollen-press of the other, and the whole compressed as the joint is closed and squeezed up into the pollen basket on the outside of the leg, where it is held in place by a stout bristle. Bumble-bees and honey bees are thus beautifully designed honey and pollen harvesters.

Not all flowers are pollinated by insects. It will be remembered that many of our trees and shrubs like hazel, as well as the grasses, are wind-pollinated. In these there is no advantage to be gained from large, showy petals, and the male flowers are arranged in long, pendulous catkins, or else the anthers hang out on long, flexible filaments so that the merest shiver in the breeze is enough to set off a cloud of pollen. Wind pollination is much more of a hit-and-miss business than insect pollination, and much more pollen has to be produced to ensure the chance landing of a pollen grain on to the stigma of another flower. It is only really effective where large numbers of individual plants grow quite close together.

One such plant is the stinging nettle, which, like the dog's mercury, grows in extensive patches. Like the dog's mercury and the red campion, it is also one of our few herbaceous species in which the male and female flowers are borne on separate plants. The nettle always grows best in the more fertile soils, especially where there is a plentiful supply of phosphorus, which explains its abundance around farms, abandoned habitations, sites of fires and in the rich soils of ditches and stream banks. Nettle beds support an astonishingly rich insect population, which may seem surprising in view of their apparently effective deterrent.

The caterpillars of two of our commonest and most beautiful butterflies, the peacock and the small tortoiseshell, feed on stinging nettles, and we were fortunate enough to come across a 'nest' of peacock caterpillars on the nettles in the hedge ditch running from the Sussex hedge to Pinehurst Farm. Only five of our butterflies lay their eggs in large clutches and have caterpillars that remain together in large batches until they are mature. It was easy to trace

Bumblebee with pollen sac on hind leg

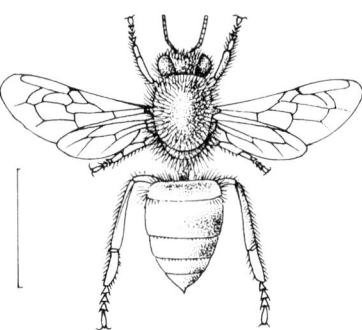

Honey bee

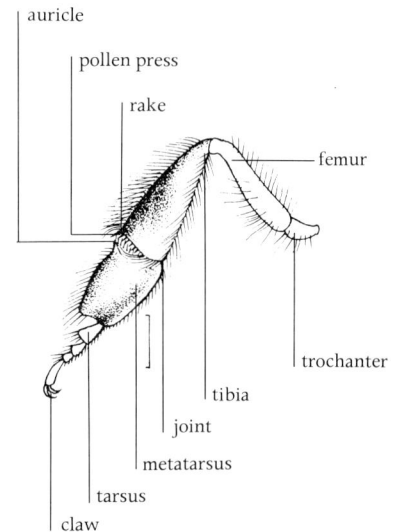

auricle
pollen press
rake
femur
trochanter
tibia
joint
metatarsus
tarsus
claw

Hind leg of bee

Stinging nettle Urtica dioica *showing male catkins*

the route along the hedge that the caterpillars had taken as they moved from one nettle plant to the next, stripping the leaves as they went.

The caterpillars are jet black in colour, lightly peppered all over with tiny white dots, and a colony of several hundred are very conspicuous indeed. On the face of it, this behaviour would not appear to provide much protection from predators – quite the opposite, in fact. However, the peacock caterpillars are also generously armed with stiff, black, branched spines which make them most unpalatable, and their self-advertisement has probably evolved in the same way as the warning coloration of bees and wasps.

The caterpillars will wander some distance from their food-plant to pupate, hanging head-downwards, often at some height above the ground. They emerge towards the end of the month, and will be seen flying until about mid-September, when they go into hibernation. A warm spell in March will stir them from their winter sleep, and the males soon begin defending territories, often along the edge of a tall hedge, and after mating, the females lay their eggs on the clumps of the new season's nettles.

Insects

The great variety of our hedgerow trees and shrubs provides food and cover for an enormous diversity of insect and other invertebrate life which is perhaps at its most abundant at this time of year.

Normally, the frantic and ceaseless activity going on among the branches is largely unnoticed, as most of the small animals that feed on the hedgerow shrubs are well-camouflaged in order to provide some protection from the attentions of potential predators. However, it is in fact a fairly simple matter to collect and study these invertebrate inhabitants of the hedgerow. All that is needed is a simple piece of equipment which entomologists call a 'beating-tray'. This consists of a light-coloured piece of strong fabric, such as calico, stretched across a convenient-sized frame. If the hedge is fairly tall and overgrown it is then a simple matter to hold the beating-tray beneath the selected branch, give it a sharp blow with a stick or a good, sharp shake, and everything will then be rudely dislodged and end up on the waiting tray below. If the hedge has been recently cut or laid, matters are a little more difficult, and the best plan is probably to pull a branch carefully away from the body of the hedge and then shake it over the tray. In any case, be sure not to damage the hedge in the process. The amount of life that will be revealed scuttling erratically amongst the debris on the sheet is truly astonishing.

The catch can be collected for examination with the aid of a 'pooter', a glass specimen tube or small bottle through the cork of which are inserted two glass or plastic tubes. The opening of one tube is placed close to the animal to be caught, and at the same time the other end is sharply sucked. The beast is thus 'hoovered up' into the pooter. It is important to cover the end of the tube through which you suck with a piece of muslin – or else you are likely to get a mouthful of assorted insects! Simple designs for both a beating-tray and for pooters are shown opposite.

Surprisingly, relatively few groups of insects have evolved the ability to feed on the leaves of trees and shrubs. A few moment's thought will show that these arboreal grazers have had to overcome the problems of being blown or washed off the leaf itself. In addition, the leaves of many plants have evolved devices to discourage grazing insects – like the thick covering of hairs on the hazel leaf, or the poisonous tannins that accumulate in mature oak leaves. Many of the 'mini-beasts' falling on to the beating-tray will be the caterpillars of moths and sawflies and various species of beetles and their larvae, especially leaf-beetles and weevils. All of these possess powerful jaws for chewing. Another group possesses piercing mouth-parts, and feeds on plant sap, such as the various kinds of bugs, including capsid bugs, frog-hoppers and leaf-hoppers and the familiar aphids. Naturally, the enormous numbers of these insects attract a similarly diverse array of predators, particularly spiders and, among the insects, earwigs, lacewings, beetles such as ladybirds and carnivorous bugs. Still others, like the bark-lice, feed on the microscopic plants, such as the fungi and algae that live

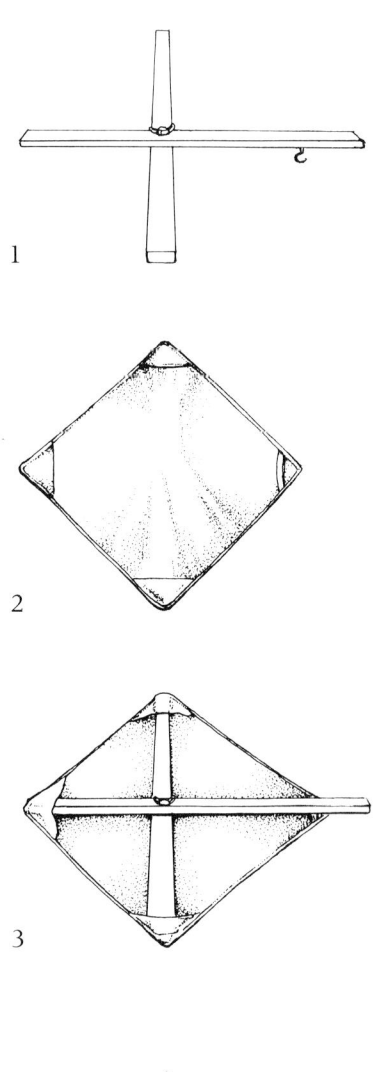

1

2

3

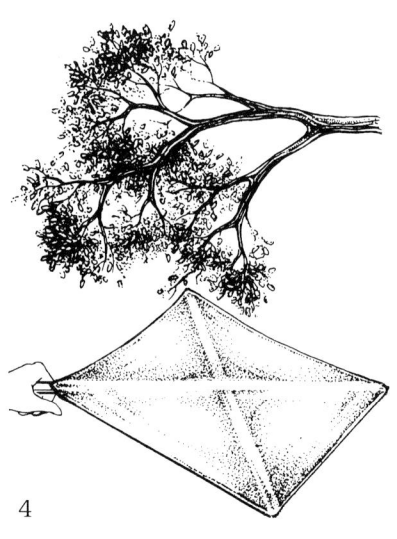

4

Simple beating tray (see left)

1 *Frame made of two strips of $2'' \times \frac{3}{4}''$ softwood (1 m long and 80 cm wide), held together in the centre by a screw and wing nut. A hook is placed on the longer piece of the frame as shown*

2 *Calico sheet 80 cm square with three sewn on corners and one loop on last corner*

3 *Assembled tray showing frame inserted into three corners and loop attached to hook*

4 *Beating tray in use*

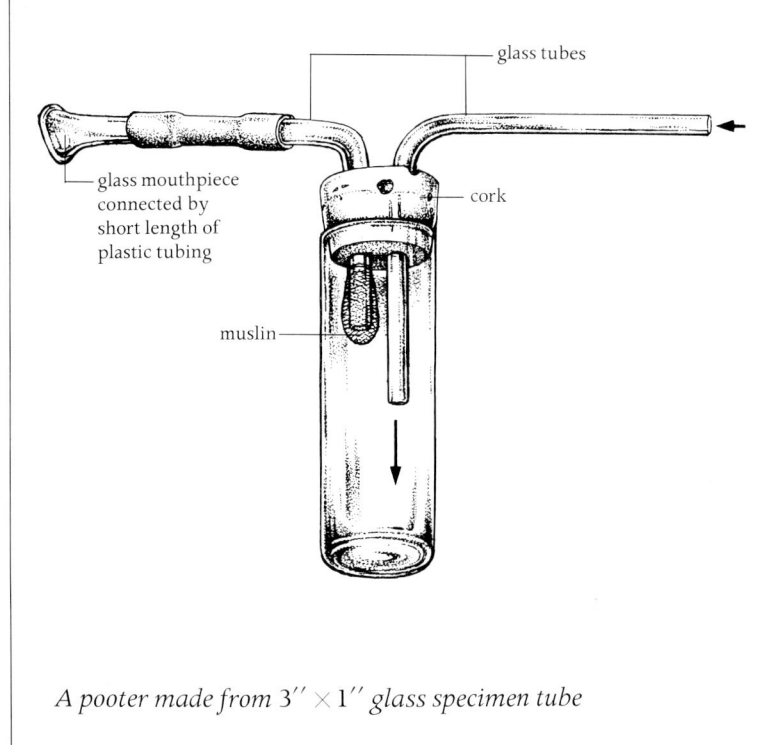

glass tubes

glass mouthpiece connected by short length of plastic tubing

cork

muslin

A pooter made from $3'' \times 1''$ glass specimen tube

Oak
Sallow/willow
Hawthorn ⎤
Sloe   ⎬ — *Rosaceae*
Apple   ⎦
Elm
Hazel
Beech
Ash
Lime
Hornbeam
Maple
Holly

on the surface of twigs and leaves. Examples of all these are shown in detail overleaf.

Interestingly enough, the leaves of some species seem to be more palatable than others. For instance, members of the rose family, the Rosaceae, such as hawthorn, sloe and crab-apple, appear to be especially favoured, whereas very few species are able to exploit the thick, waxy leaves of the holly. Not surprisingly, the oak, which is our most abundant and widespread forest tree, takes the over-all prize, notwithstanding the poisonous tannins that the leaves build up towards the end of the season. Professor Southwood has estimated that almost 300 species of bugs, beetles and moths feed on the leaves of the oak. That is not to say, of course, that every fair-sized oak can be expected to have that number of different insects on it!

Insects are found on the following list of common hedgerow trees and shrubs which are arranged in order of those with the most down to those with the least number of associated insect species (shown on the left).

The three members of the Rosaceae are clumped together near the top of the list together with the willows and sallows, another favoured group, whilst holly brings up the rear.

This great abundance of invertebrate life provides a vital source of food for the hedgerow birds, particularly at the crucial time of year when the parents have nests full of ever-hungry mouths. Even seed-eating birds feed insects to nestlings, partly to ensure that they get enough water in their diet.

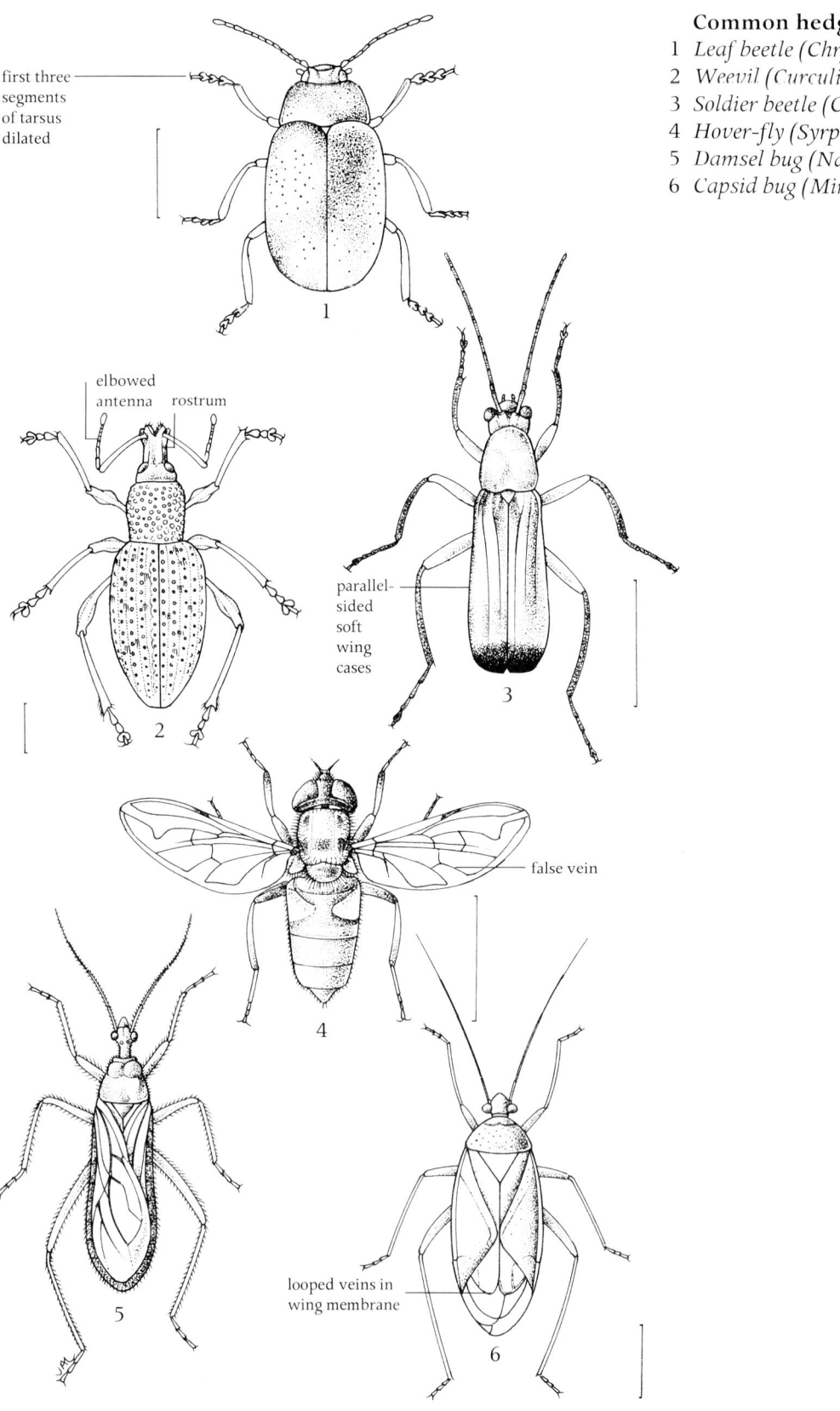

first three segments of tarsus dilated

Common hedgerow invertebra

1 *Leaf beetle (Chrysomelidae)*
2 *Weevil (Curculionidae)*
3 *Soldier beetle (Cantharidae)*
4 *Hover-fly (Syrphidae)*
5 *Damsel bug (Nabidae)*
6 *Capsid bug (Miridae)*

elbowed antenna   rostrum

parallel-sided soft wing cases

false vein

looped veins in wing membrane

7 *Lacewing (Neuroptera)*
8 *Crab spider (Thomisidae)*
9 *Winged bark louse (Psocoptera)*
10 *Aphid (Aphididae)*
11 *Sawfly larva (Hymenoptera-Symphyta)*
12 *Earwig* Forficula auricularia
13 *Moth larva (Lepidoptera)*

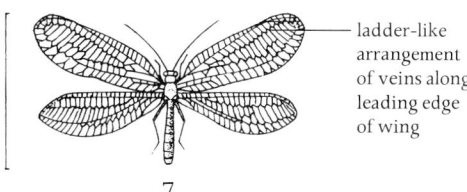

ladder-like
arrangement
of veins along
leading edge
of wing

7

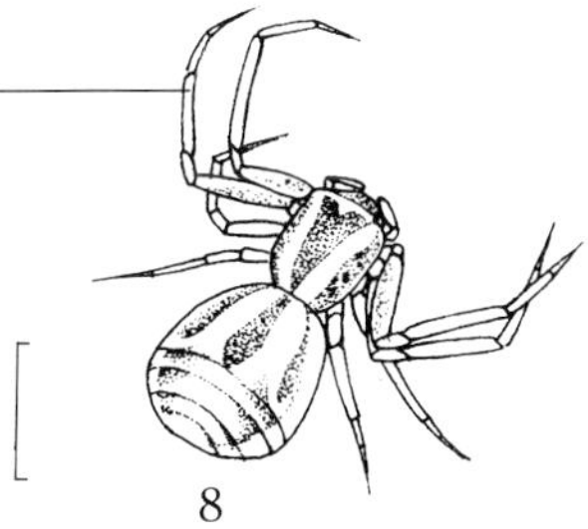

second
pair of
legs
longest

8

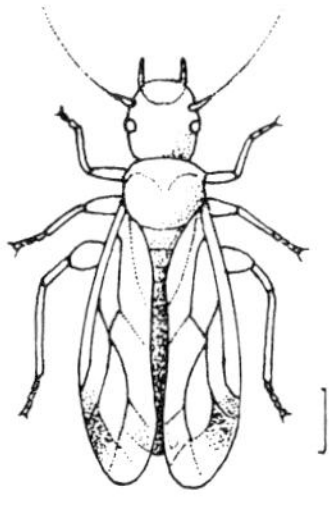

9

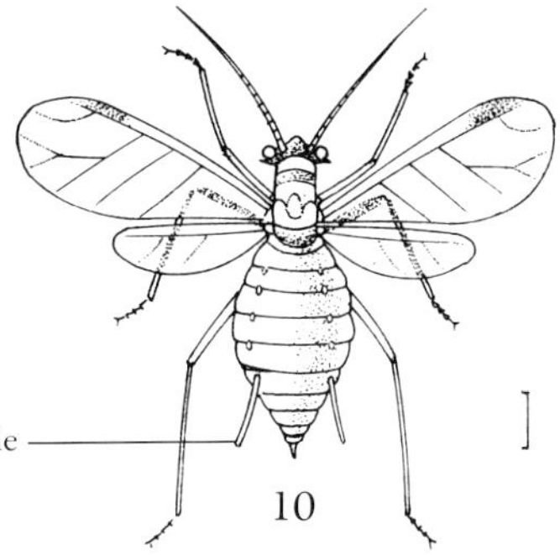

cornicle

10

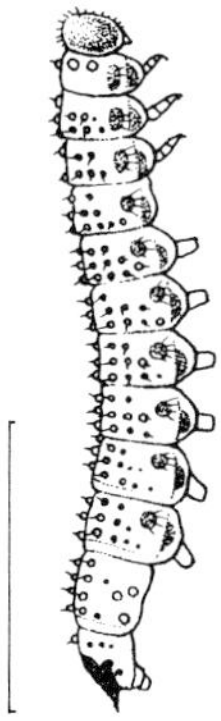

11

more than
five pairs of
abdominal
legs

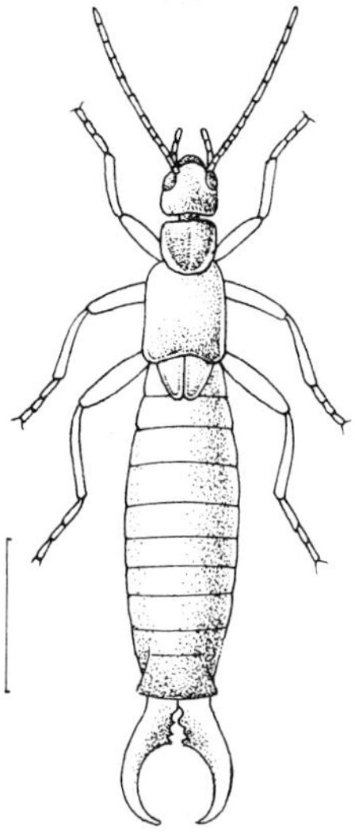

12

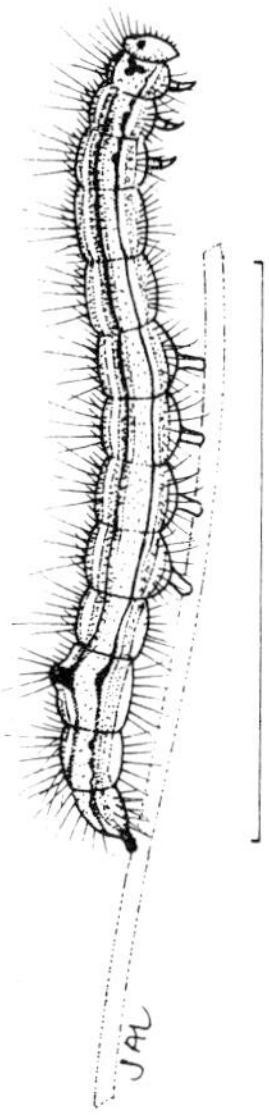

13

no more
than five
pairs of
abdominal
legs

81

July plants

The foxglove, perhaps above all others, is the flower of the fairies. There are many stories about how it got its name, the most straightforward being the resemblance of its flowers to the fingers of a glove, and from its habit of growing on disturbed ground near the earths of foxes. It was given its Latin name 'Digitalis' in the sixteenth century because of its finger-shaped flowers. It has many graphic local names such as 'granny's gloves', 'lady's thimble' and 'goblin's thimble'. In Ireland it is known as 'fairy cap' because some of the fairies there wore the flowers as headgear or gloves, and legend has it that the fairies of the moon were decked with petticoats of foxglove flowers. In Wales they are 'goblins' gloves' and 'witches' bells'.

Elves are said to hide in the bells of these magical flowers, which give mysterious powers to those who hold them, be they human or fairy: it is said that the fairies gave the corollas of this powerful plant to the foxes so that they could stalk poultry silently, or escape from man's snares as if by magic. When its tall stalks bend in the wind, it is believed that they are acknowledging a passing supernatural presence. It is unlucky to carry foxgloves on board ship.

The foxglove is one of the most poisonous plants in our flora, and yet its leaves contain a substance which is the source of one of the best-known and most widely-used medicines in heart disease: digitalis. It was in the eighteenth century that William Withering, in Shropshire, investigated the folk-use of foxglove tea for curing dropsy (which is an accumulation of watery fluid in various body tissues and cavities). He guessed that it might have wider medical uses, and, by 1799 – the year of his death – it was recognised as a valuable medicine in the treatment of heart complaints. There is a foxglove carved on Withering's tombstone in Edgbaston Old Church.

Cloth from nettles, an ancient textile plant, was made in Scotland as recently as the eighteenth century; it was used for sheets and tablecloths. The fibre makes fine, strong fabric, but since forty kg of wild nettles are needed to make just one shirt it has been superseded by more economical fibres. Nettle oil preceded paraffin, and the juice, which curdles milk, was once used instead of rennet in the production of Cheshire cheese. Pack fruit in nettles and they will preserve its bloom; grow them near root vegetables and they will improve their storing quality. A bunch of nettles hung in the larder repels flies, and nettle juice combed through the hair has been recommended as a cure for baldness.

Nettle tea, with its high vitamin C content, is renowned as a spring tonic, and street-hawkers used to cry: 'Nettles with tender shoots to cleanse the blood'. It was meant to be good for colds, and to be a good gargle.

Worn on the person in times of danger nettles drive out fear and inspire courage; they were also believed to be a protection against evil spirits. To throw nettles on the fire during a storm would protect the house from lightning, and in Yorkshire nettles

Common sorrel Rumex acetosa

82

were used to exorcise the Devil. Nettlebeds are the homes of elves, and nettle stings a protection from sorcery. A bunch of nettles hung in the dairy would protect the milk from the evil spells of witches.

In medieval symbolism the nettle stands for envy; in the Victorian 'language of flowers', for slander and cruelty, on account of its sting. To dream of being stung by nettles indicates disappointment and vexation, but to dream of gathering nettles means that someone has formed a favourable opinion of you, and if you are married that your family will be blessed with harmony and concord.

'Though you stroke a nettle ever so kindly, yet will it sting you' and 'He that handles a nettle tenderly is soonest stung' are sayings that combine practical advice with shrewd philosophy. If you are stung, however, country wisdom has a remedy:

Nettle out, dock in,
Dock remove the nettle sting.

Its common name comes from an ancient root 'ned', to twist, from its fibre. Its Latin name, *Urtica dioica*, is derived from '*urere*' to burn or sting. '*Dioica*' means dioecious: bearing male and female flowers on separate plants.

Most hedgebanks, except those in deep shade, are likely to sport their quota of vigorous growing docks. The two commonest species are the broad-leaved dock and the curled dock, both of whose names refer to the shape of the leaf. The broad-leaved dock has broad, flat leaves with a heart-shaped base, whereas the leaves of the curled dock are more or less parallel-sided with distinct wavy edges. The sorrels belong to the same genus of plants as the docks, *Rumex*, and differ in that the base of the leaf blade has two lobes.

The common sorrel is very distinctive at this time of year with its slender, rusty-red fruit-stalks standing out among the hedgebank grasses. Sorrel gets its Latin name, *Rumex acetosa*, from the sharp taste of the leaves – '*acetum*' meaning vinegar. Its local names include 'bread and cheese', 'London green sauce', 'soldiers' and 'sour grass'. *Rumex* is a kind of spear and is descriptive of sorrel's long, pointed leaves. The juice of the leaves removes stains from the hands and is said to take rust marks out of linen. A decoction of the leaves was used in folk medicine to relieve fever.

The cow parsley and rough chervil of the earlier part of the year have now been replaced by the upright hedge parsley, an annual species that does look very similar indeed to cow parsley. Perhaps the simplest way of distinguishing the two is that cow parsley has hollow stems, whereas the stems of the upright hedge parsley are solid. In the southern part of the country, especially on the more fertile or lime-rich soils, a yellow-flowered umbellifer, the wild parsnip, is now in full bloom, also attracting a mass of insects. Great care should be exercised if this plant is picked, as the smallest amount of sap on the skin is liable to produce nasty blisters when exposed to the light.

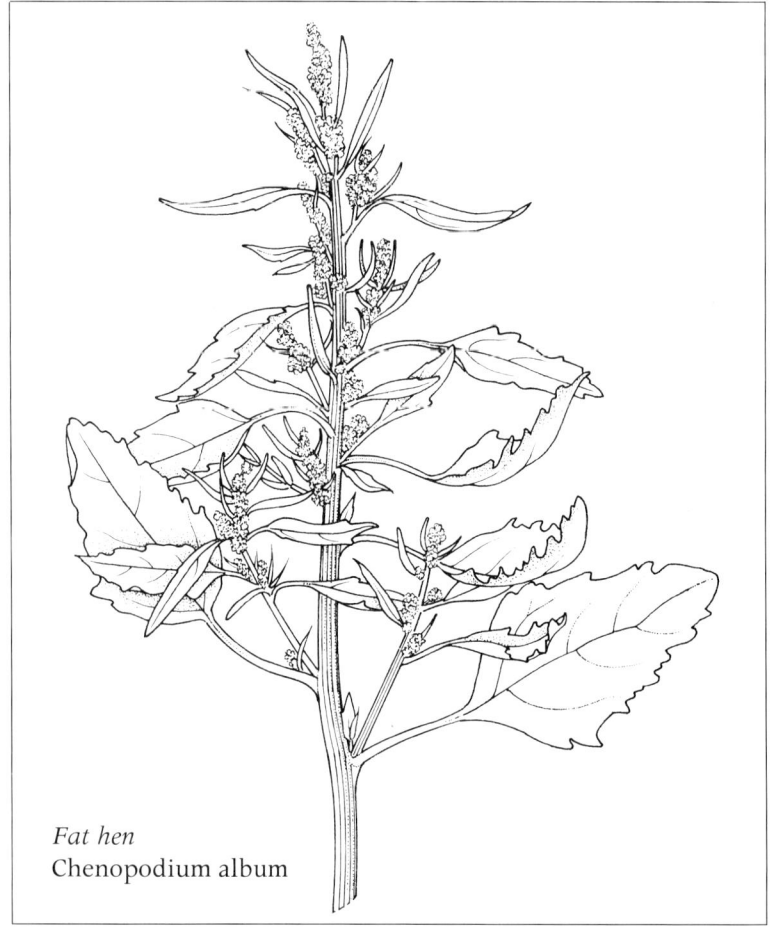

Fat hen
Chenopodium album

Farm hedgebanks frequently acquire immigrants from the neighbouring fields, usually weed species like charlock, field convolvulus and fat-hen or simply plants of waste ground like horse-radish and hedge mustard. Many of these are midsummer-flowering and so are most in evidence at this time of year.

Fat-hen grew so profusely in Anglo-Saxon times that settlements were named after it. Its Old English name was *'melde'* – hence Milden, Melbourn and Mildenhall. Our ancestors made flour from the seeds, which were harvested, dried and ground up; from this flour they made bread, cakes and gruel. It tastes rather like buckwheat and is also good raw.

It has various derogatory local names, such as 'dirty Dick', 'muckweed' (from its habit of growing on waste ground), and 'lambs' quarters', but it is also known as 'all-good'. It is a food rich in calcium, iron and vitamins B_1 and B_2, and has more iron and protein than either raw cabbage or spinach. It was ousted as a common vegetable by the introduction of cultivated spinach, to which it is related; it is also a relative of beetroot and sugar-beet.

It is called 'fat' because of its fatty seeds, which were among the first food grains of Western Europe. Commonly known as 'goosefoot' (which is what its Latin name *Chenopodium* means), it got

Yarrow Achillea millefolium

that name from its leaf shape. It produces a red to golden-red dye.

White dead-nettle is so-called because its leaves do not sting, but it is also called 'Adam and Eve' because when you look at the flowers upside down, the golden and black stamens lying side by side are like two sleeping figures in a white silk bed. Other local names include 'archangel', 'bee-nettle', 'bumble-bee flower', and 'Adam and Eve in the Bower'.

White dead-nettle tea sweetened with honey is a country cure for chills, and children have found that the dried stems, which are hollow, make good whistles.

Yarrow's Latin name, *Achillea millefolium*, commemorates Achilles, who discovered its staunching powers when he applied it to his soldiers' wounds at the siege of Troy. Yarrow is from an Anglo-Saxon word meaning 'healer', and the plant has an ancient reputation as a cure-all because of its antiseptic qualities, and also as possessing magical powers against enchantment. Its many local names include 'carpenter's weed' (from its ability to stem the flow of blood from wounds inflicted by sharp tools), 'woundwort', 'milfoil' (from the numerous divisions of the leaves), 'soldiers' woundwort', 'nosebleed', 'sneezewort' and 'yarroway'. The fresh leaf is said to alleviate toothache, and has been used in the brewing of beer instead of hops. It has been claimed that it can stop nosebleeds, and in Ireland they hang it up in the house on St John's Eve to avert illness. It is worn on the person for protection, and it would also protect a baby from harm if tied to its cradle. Yarrow tea is said to be good for colds and rheumatism, and the leaves produce a yellow dye.

Yarrow was often used in bridal wreaths and garlands – hence another of its local names, 'Venus-tree'; eaten at a wedding, it meant that the couple would love each other for at least seven years. From time immemorial it has been used in witches' incantations and for casting spells, and yet yarrow strewn on the threshold kept witches away. It also possessed powers of divination:

Thou pretty herb of Venus tree
Thy true name is yarrow,
Now who my bosom friend must be
Pray tell thou me tomorrow.

To dream of yarrow means that you will shortly hear of something that will give you great pleasure. The counting of forty-nine dried yarrow stalks is the traditional method of divining the *I Ching*, the ancient Chinese *Book of Changes*.

The strawberry is at once the fruit of Venus and yet dedicated to the Virgin Mary. In the nineteenth-century 'language of flowers' it stood for perfection, and in medieval symbolism it represented the fruits of righteousness. The leaves used to be recommended in baths for those who suffer from 'Greivous aches and paynes of the hyppes'. The juice of wild strawberries can be used as a toning complexion wash.

Sorrel Squares
Nettle Beer
Avocado with Horse-radish Dressing
Fat-hen Pancakes with
    Buttered Dead-nettle
Pasta and Yarrow Salad
Wild Strawberries and Cream
(if you're lucky enough)
Sorrel with Eggs

Sorrel Squares

a small bunch of sorrel leaves
40 g/1½ oz butter
parsley
cream
1 egg yolk
1 egg and 1 egg white
salt and pepper

Wash and chop the sorrel leaves and cook them in the butter with some chopped parsley until soft. Add a little cream and simmer for a minute or two. Stir into the lightly beaten egg yolk. Make an omelette with the egg and egg white beaten together with some cream, salt and pepper. Cook it slowly so that the bottom browns, and put the sorrel mixture in the centre to set. Turn the edges of the omelette over the top of it, turn, and cook through until well set. Leave to cool and cut into squares when cold. Serve on little squares of fried bread.

Nettle Beer
(See notes on wine-making, page 104)

1 kg/2 lb young nettles, washed
2 lemons
4·5 litres/1 gallon water
450 g/1 lb demerara sugar
25 g/1 oz cream of tartar
20 g/¾ oz brewers' yeast

Put the washed nettles in a saucepan with the thinly pared rind of the lemons. Pour on the water and bring to the boil. Simmer for fifteen to twenty minutes. Strain on to the sugar and cream of tartar, stir well, then add the juice of the lemons and the yeast mixed with a little of the liquid. Cover and keep in a warm place for three days. Remove to a cool place for two days, then strain and bottle in screw-top bottles. Keep for at least a week before drinking: it will take that time to clear.

Avocado with Horse-radish Dressing

a piece of horse-radish root
2 tbs single cream per person
salt and pepper
mixed herbs
$\frac{1}{2}$ avocado per person

Grate a little piece of horse-radish root and add to the cream. Season to taste with salt and pepper and add some finely chopped wild herbs such as wild chives, salad burnet, spearmint, etc. Fill the cavities of halved, ripe avocados and serve chilled.

Fat-hen Pancakes

For the pancake batter
(it will make about eight
    pancakes)
just under 275 ml/$\frac{1}{2}$ pint
    milk and water mixed
2 eggs
$\frac{1}{4}$ teaspoon salt
110 g/4 oz sifted flour
50 g/2 oz melted butter

Liquidise the water and milk with the eggs and salt. Then add the flour and the butter and blend for one minute longer. Cover and chill for two hours. Make the pancakes in the normal way.
For the filling
Pick a large bunch of fat-hen, wash it well and cook it in its own water, with salt, until tender. Drain, and discard any woody stalks, and chop it. Add to 275 ml/$\frac{1}{2}$ pint béchamel sauce and mix thoroughly. Season to taste.

In the centre of each pancake put two large tbs of the mixture and roll the pancake up. Place in a buttered baking dish, sprinkle with grated Parmesan cheese and dot with butter.

Bake at 375°F, 190°C, or gas mark 5 for fifteen minutes.

Buttered Dead-nettle

Pick a large bunch of white dead-nettles. Strip off the leaves and wash them. Cook them in their own water as you would for spinach, adding a little butter and salt, for about ten to fifteen minutes. Stir from time to time until the leaves are tender and have cooked right down. Season with pepper and serve with butter.

Pasta and Yarrow Salad

Serves 4

225 g/8 oz pasta shapes
1 large handful of
    young yarrow leaves
garlic
salt and pepper
olive oil

Boil the pasta shapes '*al dente*', then drain them and leave to cool. Wash and chop the yarrow leaves and crush the garlic (use one large clove or two small ones). Add to the pasta shapes with a generous seasoning of salt and pepper, and toss in olive oil until the pasta shapes are well coated. Leave for a few hours before serving to allow all the flavours to penetrate the pasta.

Sorrel with Eggs

a handful of sorrel leaves
40 g/1$\frac{1}{2}$ oz butter
25 g/1 oz flour
a little milk
salt, pepper and nutmeg
2 hard-boiled eggs

Wash the sorrel leaves, and cook them in the butter until they are brown and wilted. Add the flour and a little top of the milk until the mixture thickens. Season with salt, pepper and nutmeg. Put in the centre of a dish. Quarter the hard-boiled eggs and place them around the edge with the yolks facing outwards. Serve cold with fresh wholemeal bread and butter.

August

August, the traditional holiday month for most people, is a busy time for the farmer as he works long hours to bring in his harvest. You can hear the muffled roar of combines in the cornfields, and the incessant thump of balers on the farm. The burning of straw in the stubble fields leaves them charred and blackened, and if this is done without proper care, it can seriously scorch and damage the surrounding hedges.

1 *Hogweed*
2 *Honeysuckle*
3 *Bramble*
4 *Hedge Woundwort*
5 *Herb-Robert*
6 *Robin's Pincushion*

Dog-days, the first two weeks in August, often bring sultry, hot weather. They are named after the bright Dog-star Sirius, which during this time rises and sets with the sun. Settled weather on St Bartholemew's Day, the twenty-fourth, promises a fine autumn: there is a country saying that

If St Bartholemew's Day be bright and clear
Then a prosperous autumn comes that year.

Butterflies

Few sights enhance the pleasure of a summer's walk more than a host of butterflies fluttering before one or feeding, wings outstretched, on the hedgerow blossoms.

Above left
Brimstone on hogweed
Above right
Comma
Centre left
Speckled wood
Centre right
Hedge brown
Below left
Peacock butterfly pupa
Below right
Peacock butterfly

The peacock caterpillars that last month were feeding communally on the stinging nettles have now pupated, and the adult will be on the wing during the month. However, by far the most numerous butterflies of the hedgerow at this time of year are the various species of 'brown'. As their name implies, all the European members of this family, with the exception of the marbled white, are usually of a brownish background colour, and all have a characteristic eye-spot or group of eye-spots towards the tip of the forewing. Some also possess spots on the hindwings. The spindle-shaped caterpillars feed on grasses, and most pupate head-downwards suspended by tail-hooks from the food-plant.

If the most characteristic butterfly of the hedgerow in spring is the orange-tip, its place is taken in late summer by the hedge-brown or gatekeeper, names that amply confirm its wayside association in the minds of the early naturalists. In 1717 James Petiver published the first book wholly devoted to British butterflies, *Papilionum Britanniae*, and in it he appropriately and delightfully referred to the hedge-brown as the 'hedge-eye'.

The butterfly hatches towards the end of July and is on the wing well into September. It is particularly attracted to bramble flowers. The sexes are easily distinguishable, as the male is smaller than the female and has a diagonal dark brown band in the middle of the forewings. The eggs, which are laid singly on the leaves of grasses, hatch during September, and the young caterpillars go into hibernation in October. Feeding is resumed the following spring, and the larva finally pupates, suspended head downwards from a stem or blade of grass. The caterpillars, which are a greenish-grey colour, spend the daytime concealed close to the ground and emerge to feed only after dark. This highly effective strategy isolated, well-camouflaged individuals evading predation by venturing forth to feed only at night is in direct contrast to that adopted by the peacock and described in the last chapter. The remarkably strong association of this butterfly with hedge-rows inevitably raises the question of the nature of its pre-agricultural landscape habitat. The most likely explanation is that it was originally at home around the edges of the larger forest clearings, and indeed it can still be found flying along grassy woodland rides. Like many of our butterflies, the hedge-brown is most frequent in the South and is not to be found at all north of the border.

Several other members of the 'brown' family of the butterflies, or the Satyridae, are frequent along the hedgerows at this time of year. Unlike the hedge-brown, the wall produces two generations in a season. The first is on the wing in May and June, and the second appears at the earliest in July. As its name implies, it is particularly fond of basking in full sunshine on walls and dry hedgebanks. In contrast, the speckled wood is primarily a butterfly of small woodland clearings where the males defend their sun-spot territories, but they are also frequently met along shaded hedgerows. Like the wall, the speckled wood produces two generations a year, the first sometimes emerging as early as April. The

Life cycle of the hedge brown or gatekeeper butterfly

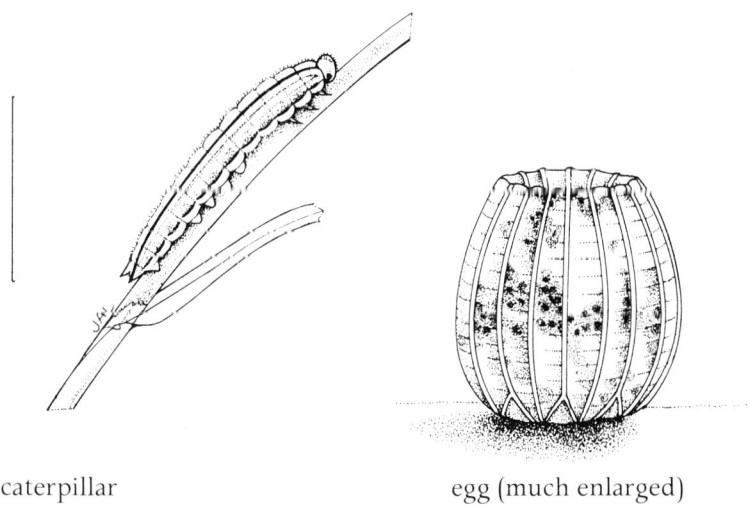

caterpillar   egg (much enlarged)

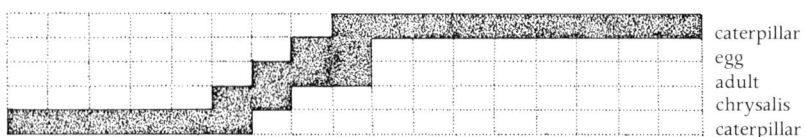

																	caterpillar
																	egg
																	adult
																	chrysalis
																	caterpillar

Jan Feb Mar Ap May Jun Jul Aug Sep Oct Nov Dec Jan Feb Mar Ap May

seasonal life cycle

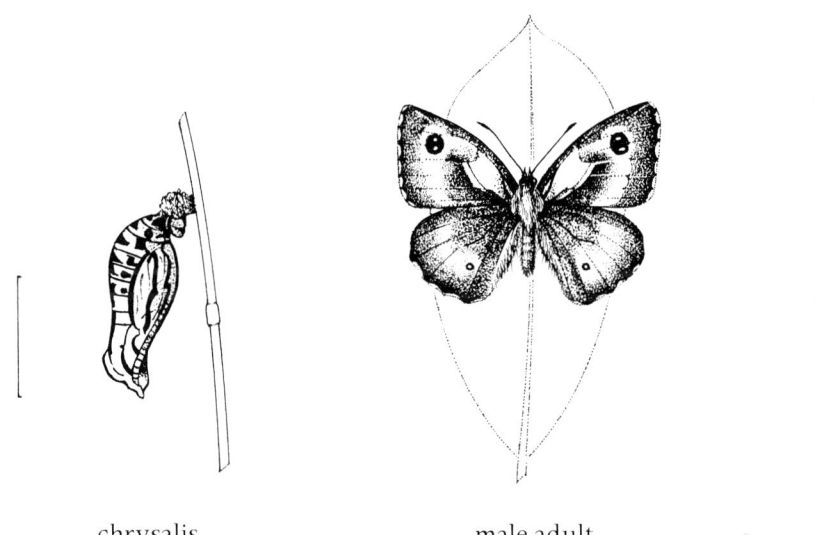

chrysalis   male adult

92

Life cycle of the comma butterfly

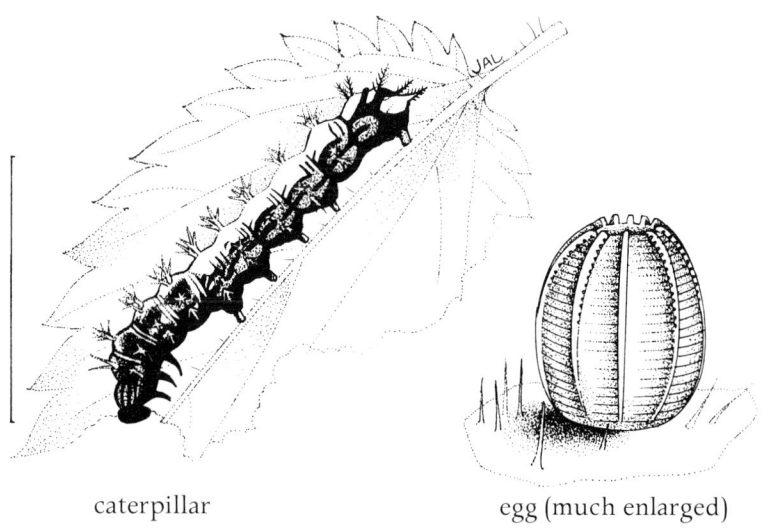

caterpillar   egg (much enlarged)

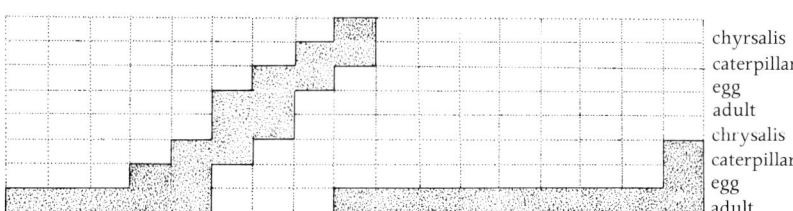

| | Jan | Feb | Mar | Ap | May | Jun | Jul | Aug | Sep | Oct | Nov | Dec | Jan | Feb | Mar | Ap | May |

seasonal life cycle

chrysalis   adult

93

fourth member of the family to frequent the hedgerow is arguably Britain's most widely distributed butterfly, the meadow brown. Truly a butterfly of pastures and rough grassland, it is often to be seen flitting about among the tall grasses bordering the hedgerows along roadsides and field borders. Like other members of the family, the males tend to be smaller and are darker than the females. Indeed, the great Swedish naturalist, Linnaeus, found them to be so distinct that he originally named the two sexes as separate species!

Two other butterflies may be met along the hedgerow at this time of year. The beautiful brimstone, sulphur-yellow in colour, is one of our few butterflies that pass the winter as the adult. It emerges from the pupa at the end of July or the beginning of August when it can be seen flying along lanes and hedgerows where the food-plants of the caterpillar, buckthorn and alder-buckthorn occur. The male is a darker, more intense yellow than the female.

Lastly, late summer is the time when one of our most spectacular and fascinating butterflies should be watched for. At the beginning of the century the comma was a scarce butterfly, more or less confined to the Welsh border counties. Happily, today it can now be expected over a large area of southern England. The margins of the wings are irregularly scalloped, and the upper surfaces are a deep rich tawny, speckled with irregular black dots. The under-sides of the wings are dark brown, and in the middle of the hind wing is carried the distinctive white 'comma' from which the insect derives both its English and Latin name, *Polygonia c-album*. The adult caterpillar is blackish and covered by spines, but there is a broad, white patch on the back which presumably either serves to break up the outline of the beast, so confusing potential predators, or alternatively mimics a bird dropping – equally off-putting! It feeds on nettles, hops or, as we were lucky enough to find it in a Gloucestershire hedgerow, on elm.

Herb Robert

Herb Robert is a flower which, as well as growing in the hedge-rows, also loves the walls and roofs of our houses, and it is almost as familiar to us as a robin: so perhaps it got its name from our native household goblin, Robin Goodfellow. Robin is a diminutive of Robert, and herb Robert is linked in folklore to goblins and magic. Alternatively its name could come from its colour: *herba rubra*, or red plant. Another theory is that it is named after Robin Hood, the thirteenth-century outlaw and folk hero. Or did St Robert, who founded the Cistercian Order, cure the plague with this plant and so have it named after him? Whichever is the case, herb Robert is certainly a powerful medicinal plant; it was held in high repute in the Middle Ages and is still employed in folk medicine for staunching wounds, curing ulcers and relieving inflammations.

Its hairy stems turn red in bright sunlight, and when its flowers hang downwards it is a sign of impending bad weather. It is also called 'stinking Bob' because it has an unpleasant smell and is quite a good insect repellent. Its other local names include 'Robin flower', 'red Robin', 'birds' eyes', 'poor Robin' and 'Robin Hood'.

Herb Robert is a species of geranium or cranesbill (so called because its fruits look like the bill of a crane). In the Middle Ages people grew it in their gardens and loved it as much as we love our cultivated geraniums today.

Plant galls

The abnormal growths on plants known as galls are among the most frequent objects to excite attention and question in the countryside. Many are at their best and most conspicuous in late summer, and several are particularly characteristic of hedgerows. Galls are caused by a variety of different organisms which stimulate the affected part of the host plant to abnormal development. Most are caused by animals but a few by fungi. The most usual gall-forming animals are eelworms, mites and, among the insects, gall-wasps, gall-midges, aphids and saw-flies.

Perhaps the most bizarre of all hedgerow galls is the 'Robin's pincushion', also known as 'briar balls', 'moss-gall' or 'bedeguar gall', commonly found on both the dog-rose and field rose. The common country name is evidence that it has long been the focus of folk-lore, as 'Robin' is again the house goblin and woodland sprite, Robin Goodfellow. The origin of the unusual word 'bedeguar' is more obscure, but it has been suggested that it is derived from the Persian word '*badawar*', meaning wind-borne, as to country people it must have appeared to have been conjured out of the air.

The gall is caused by a minute gall-wasp, called *Diplolepis rosae*. The female wasp lays her eggs in the buds of the rose in late spring. The young gall is greenish in colour, but as it grows throughout the summer it turns through pink to bright red, and by August the irregular, moss-like ball may be up to 7 to 10 cm in diameter. If the gall is cut in half at this time of year it will be found to contain a large number of small chambers, in each of which is a tiny white grub, the larva of the gall-wasp. They will remain in the gall over the winter, pupate and finally emerge next May. The gall-wasps can quite easily be reared by cutting the galls from the bushes at the end of the winter and storing them indoors on sand in jars covered with muslin. The only problem with this is that not all the insects that emerge will be the gall-causer. The Robin's pincushion has developed into a most complicated community of parasites and fellow-travellers. A second gall-wasp, *Periclistus*, has taken advantage of the situation by laying its eggs directly in the ready-made gall, a kind of uninvited, non-paying guest or 'inquiline'. Both the original gall-causing *Diplolepis* and the inquiline *Periclistus* are attacked by a variety of parasitic wasps. *Torymus bedeguaris* is a tiny metallic-green Chalcid wasp, the female of which has a long, bristle-like ovipositor with which she is able to pierce the gall and reach the cells containing the *Diplolepis* larvae. These are also attacked by an ichneumon wasp called

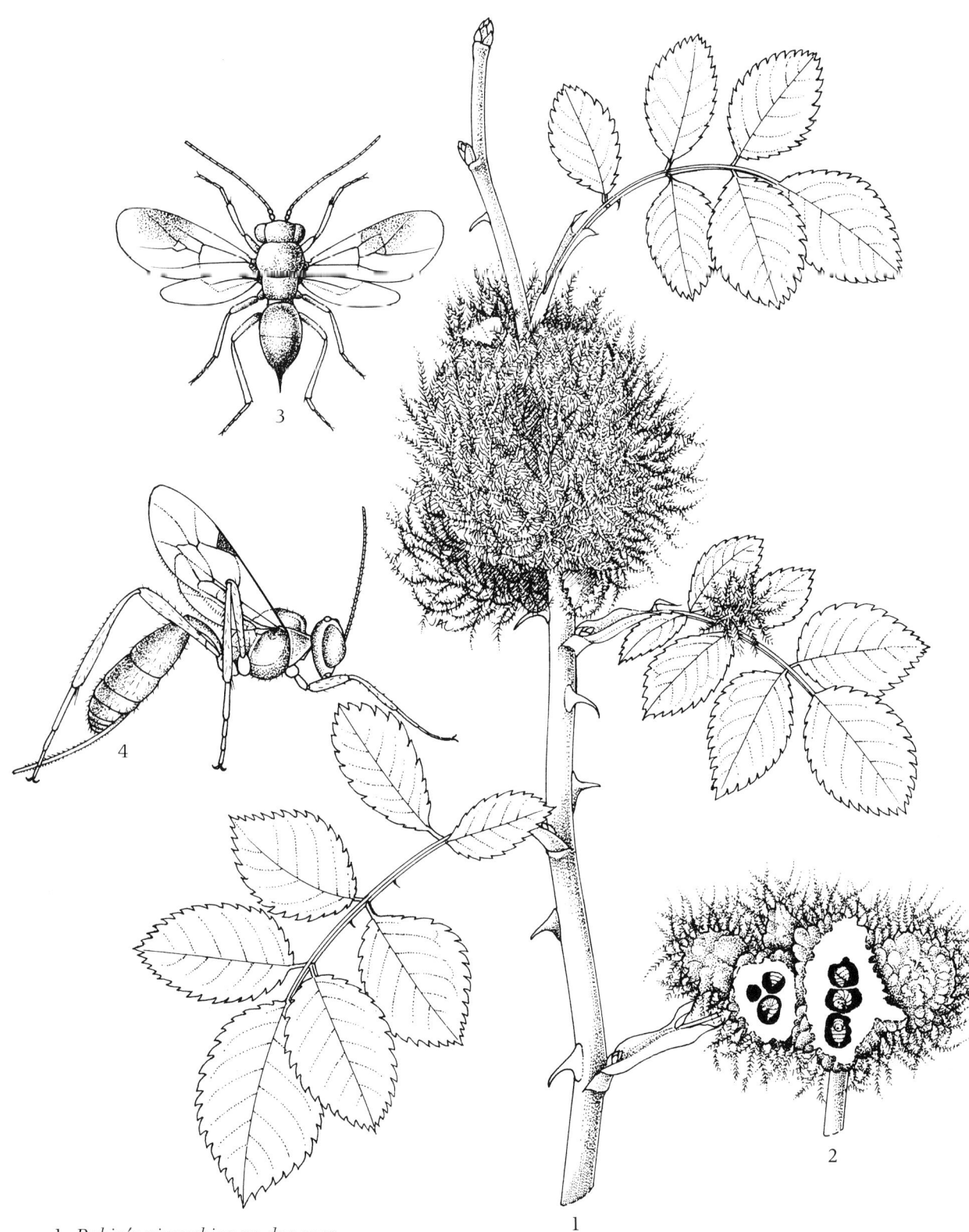

1 *Robin's pincushion on dog rose*
2 *Cross-section of gall showing wasp larvae in cells*
3 *Gall wasp* Diplolepis rosae, *the gall former (much enlarged)*
4 *Ichneumon wasp* Orthopelma mediator, *parasite (much enlarged)*

Above
Nail galls caused by the mite
Eriophyes macrorynchus *on field maple*
Above right
Marble galls caused by the gall wasp Cynips kollari *on oak*

Orthopelma and a second Chalcid called *Oligosthemus* which also infects the *Periclistus*. These three parasites are in turn parasitised by a third species of Chalcid, *Habrocytus bedeguaris*: 'Great fleas have little fleas upon their backs to bite 'em'! Just to complete the story the *Periclistus* is preyed upon by yet another Chalcid, *Eurytoma rosae*, which roams the gall from cell to cell, searching for larvae to consume. This astonishingly complex set of inter-relationships, all played out within a single Robin's pincushion, is depicted diagrammatically opposite.

Bedeguars were used by the old apothecaries for a host of doubtful remedies. A concoction of the powdered galls was supposed to cure diarrhoea in cattle, act as a diuretic and cure the colic. Even the grubs feature in one of Nicholas Culpeper's more bizarre cures: 'In the middle of the Balls are often found certain white worms, which being dried and made into powder, and some of it drunk, is found by Experience of many, to kill and drive forth the Worms of the Belly'! In some parts of the country people used to wear a Robin's pincushion around their necks to protect them from whooping cough.

Diligent searching of the hedgerow bushes and herbs will reveal a whole range of different galls. The tiny red pustules that pepper the leaves of the field maple are caused by a mite, *Eriophyes macrorynchus*. A similar species, *E. similis*, causes irregular green pustules along the leaf margins of the blackthorn. The stunted shoots and closely packed, distorted leaves of the hawthorn are evidence of the gall-midge, *Dasyneura crataegi*, whilst a related species, *D. fraxini*, causes cigar-shaped galls on the midrib or the under-surface of ash leaflets. Midge galls can also be found on the leaves of ground ivy and in the shoot tips of the germander speedwell.

By far the largest number of different galls are found on the oak, such as the familiar oak-apple, marble galls and the currant and spangle galls caused by different generations of the same species of gall-wasp.

97

The various blotches and squiggles that occur on the leaves of a whole range of hedgerow plants are yet another familiar but puzzling phenomenon. These 'leaf mines' are caused by the larvae of various insects, especially flies or small moths, which tunnel and feed between the upper and lower epidermis of the leaf. The nature of the food-plant, the shape of the mine and its position on the leaf are all characteristics of the particular species. Common examples include the serpentine mines of the moth *Nepticula aurella*, frequently seen on the leaves of the bramble. The egg is laid at the narrow end of the mine, and it widens out as the larva matures. On the other hand, the similarly shaped mines that appear on the leaves of the primrose are caused by a fly, *Phytomyza primulae*, as also are the blotch-shaped mines on the leaves of holly, caused by the related *Phytomyza ilicis*. The eggs are laid singly on the undersurface of the holly leaf in May or June, and, after hatching, the larva feeds by tunnelling along the midrib for about two months before beginning to feed on the inside of the leaf to produce the characteristic mine. The larva overwinters in the mine and finally pupates the following March before emerging as the adult of the new generation.

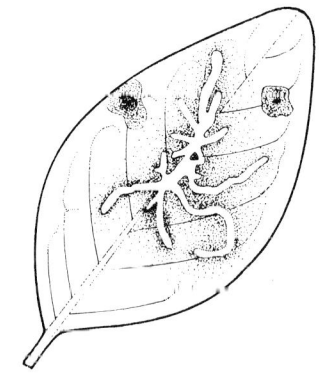

Honeysuckle leaf mined by the larva of the fly Phytomyza periclymeni

Honeysuckle leaves also are often disfigured by the mines of a closely related fly, *Phytomyza periclymeni*. Honeysuckle is one of the most beautiful of our midsummer hedgerow flowers, and its yellow colour and long corolla tubes are typical of moth-pollinated plants. Our honeysuckle has rather a restricted geographical distribution, being more or less confined to Western Europe. However, it is common in woodlands and hedges throughout Britain but often fails to flower when growing in heavy shade.

A flower of love and poetry, for Chaucer the emblem of fidelity, honeysuckle was traditionally presented by ardent lovers to their sweethearts as a demonstration of their passions. In the Victorian 'language of flowers' it stood for generous and devoted affection. Like a mountain goat it leaps and binds beyond the places where mortals dare to climb, so it has been called 'goat's leaf'. Garlands of honeysuckle were used by witches in spells to cure sick children, and also to avert evil spirits on May Day. In Scotland farmers used to place branches of honeysuckle in their cowsheds on 2 May, to keep their cattle from being bewitched.

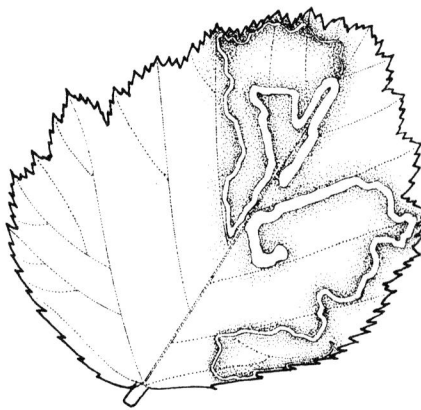

Bramble leaflet mined by the larva of the moth Nepticula aurella

Honeysuckle gets its name from the sweet nectar in its corollas which makes them so good to suck. Its berries, on the other hand, are poisonous. It is commonly known as woodbine, too, from its habit of binding tightly around trees. Other local names include 'hold-me-tight', 'eglantine', 'fairy trumpets', 'honeybind', and 'withywind'. Pepys called it the 'trumpet flower', whose 'ivory bugles blow scent instead of sound'.

One of the most striking features of the late summer hedgerow is again a fine display of tall, robust umbellifers – this time of the hogweed, which will remain in flower for the rest of the season. The broad white umbels attract a quite staggering variety of insects at this time of the year, including the distinctive orange and black soldier beetles, *Rhagonycha fulva*, that we mentioned in July. The

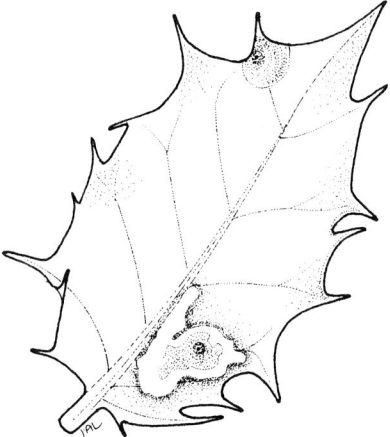

Holly leaf mined by the larva of the fly Phytomyza ilicis

98

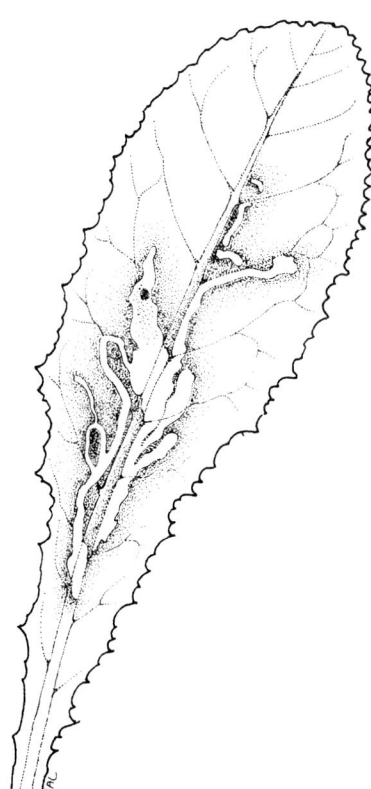

Primrose leaf mined by the larva of the fly Phytomyza primulae

beetles, which are just under 1 cm long, are equally abundant on other flowers such as ragwort, feeding on pollen and smaller insects. The larvae are wholly carnivorous, living in the litter at the base of the hedgerow grasses. Another, smaller soldier beetle, *Malachius bipustulatus*, should also be looked for at this time of year. It is a beautiful metallic green in colour with red tips to the wing cases.

The hogweed delights in such names as 'cow clogweed', 'humpy-scrumples', 'humperscrump', 'madnep' and 'pig's flop'. The root is indeed used as fodder, and it was not an uncommon sight to see country people harvesting cow parsnip or pigweed, as they called it, for their pigs.

The hollow stems are deeply furrowed and are popular with children as pea-shooters. But beware of the exotic giant hogweed with its purple blotched stem which grows up to 10 feet tall: touching it can cause blistering and unpleasant skin irritations.

Hogweed is said to have healing properties which were discovered by Heracles, the god of strength – hence its Latin name '*Heracleum*'. It relieves hypertension, was once used to treat epilepsy, and still holds a place in homoeopathic medicine.

The folklore of white bryony is contained in its large, tuberous root: it became the home-grown mandrake of medieval quacks who invested it with false magical powers – hence one of its local names, 'English mandrake'. Modelled into the shapes of men and women, it was used by witches to make ugly images of those upon whom they intended to exercise their magic. The root may be made to grow into any shape by putting it, when young, into an earthenware mould, and its twisted forms used to be sold as charms and curiosities, and even consulted as oracles. Human shapes made from bryony roots used to be hung up as herbalists' shop signs – but this was pure skulduggery since the imposters were claiming it to be mandrake, which of course it was not.

A member of the gourd family and commonly known as 'wild vine', white bryony was the white vine of ancient medicine. Its mildly poisonous berries have been used as a purgative, and infusions are recommended for catarrhal and rheumatic complaints, and to relieve coughs in pleurisy. It is also said to have aphrodisiac properties. Its Latin name, *Bryonia dioica*, comes from a Greek root meaning to swell, or grow luxuriantly. '*Dioica*' means 'dioecious', bearing male and female flowers on separate plants.

Hedge woundwort was given its common name from the use of its leaves in healing cuts and wounds. Like its cousin marsh woundwort, its leaves were made into an ointment or used as a poultice, because they contain – as is now known – a vaporisable antiseptic oil. Its Latin name, *Stachys sylvatica*, derives, by association, from the Greek for a spike, which describes its flowerhead; *sylvatica* means 'of the woods'. Gerard called marsh woundwort 'clown's all-heal' because of his personal dramatic (and successful) attempts to heal the grievous wounds of two of his patients, which he vividly describes in gory detail in his *Herball* of 1598.

Dating hedges

The history of a hedge has a profound effect on the richness of the plant and animal life that it supports. Generally speaking, it seems to be the case that the older the habitat, the larger will be the number of animals and plants associated with it, so it is hardly surprising that this appears to hold true for hedges.

It will be recalled that hedges started life in a variety of different ways. The ancient woodland relic hedges might quite easily have an uninterrupted link with the forest of the pre-agricultural landscape. At the other end of the scale, many enclosure hedges from the sixteenth century onwards were originally planted with single species and have since been progressively colonised by other species of shrubs. Between these extremes are the hedges that have been planted at various times from the mediaeval period onwards with several different species, either collected from the surrounding waste or more recently from nursery-reared stock. However, whatever the detailed history of the hedge, it is likely to be true that a greater variety of shrubs will be found in the older hedges. In the mid-1960s, during the course of some research into the variation in shrub diversity between different hedges, Dr Max Hooper became puzzled by the apparent lack of correlation between differences in soil, climate and management and the composition of the hedges that he was looking at. The clue came when he compared the species composition with the age of some Devon hedges that could be dated from documentary evidence. From this Dr Hooper concluded that it appeared possible that one additional species of shrub colonised a hedge every hundred years. Later he was able to confirm this from similar work in other parts of the country. For instance, in a small area of clay uplands on the Huntingdonshire–Northamptonshire border, age accounts for eighty-five per cent of the variation between hedges, leaving other factors such as small differences in soil properties and management history to account for the remaining fifteen per cent. Furthermore, by pooling the data from almost 230 hedges from Devon, East Anglia and the east Midlands he was able to calculate an equation representing the way in which the number of shrub species in a thirty-yard stretch of hedge varies with its age:

Age of hedge $=$ (number of species $\times 110$) $+$ 30 years

In other words, a hedge with two species in a thirty-yard stretch would be about 250 years old, and one with six species about 690 years old. This is only a rough approximation, and the calculation could be 200 years out on either side. However, it does provide a rule of thumb estimate of about one shrub species for every hundred years. The technique also has the virtue of simplicity, providing that a number of simple procedural rules are observed.

First, it must be remembered that the ends of a hedge are likely to be atypical, especially where it joins on to a wood or copse. It is advisable to pace out a set distance from the end of the hedge – say,

Left

White bryony Bryonia dioica. *The only British member of the Cucurbitaceae, the cucumber family and a common hedgerow plant of chalky soil in southern England*

ten paces – and record the shrubs in the next thirty paces, as this also ensures that you don't succumb to the temptation of deliberately choosing the most species-rich and interesting part of the hedge to sample! The accuracy of the method does not warrant the use of a tape-measure. It is a good idea to sample three or four consecutive thirty-yard strips in each hedge and calculate the mean. Only one side of the hedge is recorded, and all species of trees and shrubs are counted, with the exception of woody climbers and brambles. Honeysuckle, traveller's joy and ivy, therefore, wouldn't count, and all species of rose should be lumped together as one. Ideally, ten or more hedges that can also be dated from documentary evidence should be counted in any particular area in order to establish as accurate a relationship as possible between the number of shrubs and the age of the hedge, as this does vary in detail between different parts of the country.

It will be remembered that the 1771 map revealed that the Essex hedge was a farm boundary in the eighteenth century so that, assuming that the boundary was hedged (although the map doesn't actually illustrate a hedge) we concluded that the hedge must be at least 200 years old. We counted the shrubs and trees in three successive stretches of the hedge, starting ten paces in from the farm end. These gave us nine, nine and eight species, suggesting that the hedge was about 800–1,000 years old. This is a very high and interesting count on any basis, and suggests that the hedge is some 600–800 years older than the earliest documentary evidence of the boundary that we have been able to find. However, it does correlate very nicely with the assize roll reference to the thirteenth-century Hugh Spon who possibly bequeathed his name to the farm.

The historical evidence suggests that the Sussex hedge, on the other hand, might be quite a lot older, as not only is it a parish boundary but also it grows on the boundary of the ecclesiastical Saxon Manor of Malling. The species counts for this hedge gave us consecutive figures of seven, seven and eight. From this it looks as if the hedge may not be as old as the boundary, even though a medieval origin is indicated.

The results from the surveys of our two hedges serve to emphasise that one should be cautious in attempting a too precise relationship between age and number of species. However, the counts alone do point to a pre-Tudor date for both hedges, whilst the evidence from the shrubs in the Essex hedge in particular strongly indicate a much more ancient origin than the earliest map suggests. Bearing in mind the margin of error, the technique should be sufficiently sensitive to distinguish at least between a medieval and older, a Tudor or a nineteenth-century enclosure hedge. This is valuable information, as the identification of ancient hedgerows is now a vital part of countryside conservation.

Considerable evidence is now accumulating to suggest that other kinds of natural history evidence can be used in addition to the number of shrubs to throw light on the age of hedgerows. Some species of shrubs and herbs seem to be characteristic of the older hedges. For instance, we suggested in the April Chapter that hedges of hazel and maple are often older than hedges consisting almost wholly of hawthorn. This generalisation needs some modification, as the occurrence of the Midland hawthorn, described in the May Chapter, is almost always good evidence of an ancient hedge. It was the common hawthorn that was so widely planted in the new enclosure hedges of the eighteenth and nineteenth centuries. Spindle is another shrub that does not seem to occur in the more recent hedges.

Similarly, certain species of woodland herbs may also be indicative of an old hedge, especially in parts of the country like East Anglia and Lincolnshire, where woodland is generally scarce. Such plants as dog's mercury, bluebell and yellow archangel would fall into this category. Even some species of snails have been found to be more characteristic of old than new hedges in a study carried out in Shropshire. The common little woodland snail of decaying wood, *Discus rotundatus*, is one such species. There is little doubt that further research will reveal other groups of animals that can be used to assist in hedge dating. Obviously, the more evidence that can be independently accrued from different groups of plants and animals, the more reliable will biological dating become.

Making Country Wines and Cordials

April: Dandelion Wine

May: May Blossom Wine Cup

June: Elderflower Cordial

July: Nettle Beer

August: Blackberry Wine

September: Elderberry Wine

October: Sloe Gin and Bullace Vodka

Every enthusiastic wine-maker – like every cook – has a favourite method, based on personal experience of success and failure. There are a few general rules, however, that never fail, and one is cleanliness. Always make sure that your containers – be they pans or bowls or bottles – are scrupulously clean. After washing them you can use a solution of 1 Campden tablet dissolved in $\frac{1}{2}$ pint/275 ml water to sterilise them.

The second important factor in successful wine-making is temperature, the ideal for fermentation being around 70–74°F/21–24°C. You want to keep this constant, so a place next to a boiler, or in an airing cupboard, is perfect.

The third golden rule is patience. Having gone to all the trouble of making your wine, give it a chance to mature fully before drinking it: it is worth it. That goes for sloe gin too: it is far better for being a year old, if not older. White wines need less time – but usually at least six months, whereas red wines will profit from two years or more. For the impatient: nettle beer will be ready after a couple of weeks; and for the hasty, elderflower cordial and May blossom wine cup can be drunk almost as soon as they are made!

With all the recipes for hedgerow drinks in this book, be sure to wash the flowers or fruits well before embarking on the recipe: for perfectionists, a solution of a Campden tablet in water will guarantee that your ingredients are sterile as well as clean.

Blackberry Wine

2 kg/4 lb blackberries
3·5 litres/6 pints water
1 kg/2 lb sugar
1 teaspoon citric acid
1 sachet wine yeast
Campden tablets

Clean, wash and crush the fruit and pour boiling water over it in a sterilised bowl. Cover and leave to cool. Stir in the sugar, acid and yeast, replace the cover and ferment on the pulp for four days, pressing down the floating fruit twice a day.

Strain out, press dry and discard the fruit, and pour the must into a sterilised fermentation jar. Fit an airlock, and leave in a warm room until the wine is still and beginning to clear, about 2 weeks.

Syphon the clearing wine into a sterilised jar, discard the sediment and top up with cold boiled water. Add 1 Campden tablet, bung it tightly, and leave in a cool place until the wine is bright. Syphon into bottles and store for at least one year.

This makes a well-flavoured, dry, red table wine to be served at room temperature with red meats and cheese dishes.

September

The ragged noisy boy intrudes
To gather nuts that ripe and brown
As soon as shook will patter down
Thus harvest ends its busy reign
And leaves the fields their peace again
Where autumns shadows idly muse
And tinge the trees with many hues

John Clare, 'The Shepherd's Calendar'

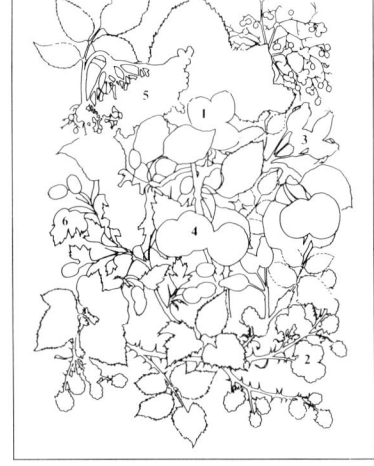

1 *Hazel*
2 *Bramble*
3 *Rose hips*
4 *Crab Apple*
5 *Elder berries*
6 *Hawthorn*

September brings early morning mists and the first heavy dews sparkling in the mellow sunlight. The colours of the countryside are changing: black bryony is clambering high into the hedges with a dramatic display of berries, some already bright red, some still green. Blackberries are ripening, and birds prepare to migrate. But although there are hints of autumn, summer lingers on so that in September we often have the best of both worlds. Many plants are still in flower: hogweed, hedge woundwort and white dead-nettle, angelica in damp hedgerows and the glorious honeysuckle. Very often at the end of the month there is an 'Indian summer', so called in the nineteenth century by North Americans because the Indians used such days to store their crops and prepare for winter.

The twenty-ninth of the month is the feast day of St Michael the Archangel: Michaelmas is one of the quarter days in England, and traditionally it was the landlord's rent-audit day. He would prepare a feast of hot roast goose for his tenants in the hall of the big house. If acorns fall on St Michael's Day there will be a bitter winter: but if the birds have not migrated by then, then winter is still a long way off – an optimistic prophecy as late stragglers of swallows and house martins are always about well into early October!

Fruits

Above all, September is the month of berries – bramble, rowan, hawthorn, black bryony and spindle, to name just a few: a hedgerow harvest which will later attract flocks of winter birds.

The blackberry has also attracted hordes of human pickers and has provided food for man since neolithic times. The bramble is not only useful for its fruit, however: its twigs are used in wickerwork, and the plant provides natural dyes, black from the young green shoots, slate-blue from the berries and orange from the roots. As a bonus, sheep leave their wool on its thorns for collecting and subsequent spinning.

The bramble has much folklore attached to it: sick children used to be passed through an arch of bramble that had rooted at both ends in order to make them better; it was also believed that this could cure rheumatism. The Greeks and Romans used blackberries to cure gout, and infusions of the leaves are still used in folk medicine to relieve sore throats and tonsillitis. Current medical researchers are investigating the possibility that the leaves contain anti-diabetic properties.

The bramble has religious associations too: was this the burning bush in which Jehovah appeared to Moses? Or perhaps the Crown of Thorns? In the Old Testament it was chosen to rule over the giants of the plant kingdom. Perhaps the best-known part of blackberry folklore, however, is the story that, as Satan was cast out of Paradise, he fell into a bramble bush and cursed it roundly. He is now said to spit (or less delicately, to urinate) on the plant on every anniversary of the Fall: hence the superstition that it is unwise to eat blackberries after Old St Michaelmas Day, 11 October, as they will be sour.

Elder is unusual among our hedgerow shrubs in its requirements for a very fertile soil. It will grow well only where the soil is rich in nitrogen and phosphorus – which is one reason why it is such a familiar sight around farmyards. It is also common around rabbit warrens, benefiting from the organic enrichment from the rabbits' droppings while the rabbits themselves leave it alone. In the hedgerow, it usually appears where the soil has been disturbed, as in the Sussex hedge where it grows all round the badger set.

It is a highly magical tree which has more folklore attached to it than any other tree of its size, and its berries are much sought after by birds and under-estimated by man. Although their pungent taste and rich texture are an acquired taste, they complement other fruits very well and, mixed with apples, blackberries or crab-apples, they make beautiful preserves and pies. They are extremely nutritious and were much used in ancient medicine – indeed, they were regarded as a cure-all. John Evelyn said that an extract from the berries was 'a catholicon against all infirmities whatever'. Gerard had a use for them in preparation: 'The seeds contained within the berries dried are good for such as have the dropsie, and such as are too fat, and would faine be leaner, if they be taken in a morning to the quantity of a dram of wine for a certain space.' However, eating the berries raw can make you sick (perhaps that was Gerard's slimming method), so, uncooked, it is best to avoid them. They give a lilac dye.

The appearance of elderberries was held to indicate the right time for sowing wheat; and country people used to beat their fruit trees and vegetables with branches of fruiting elder to impregnate them with the scent of elderberries in order to deter insects.

Wilding, scrab, gribble and scrogg could be a quartet of hobgoblins – but no, they are all local names for the wild crab-apple tree. Why 'crab'-apple? Perhaps because it is an ungainly tree which bears unrewarding fruit, so its name is derogatory. Its redeeming features are its lovely blossom in the spring, and the fact that it is host to the magical mistletoe. The wood of the crab-apple is very hard and was used in the past for wood carving and making mallets: its strength evidently came in handy too for husbands angry with their errant wives:

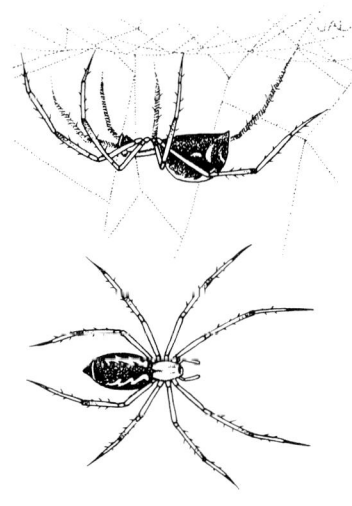

The crab of the wood is sauce
  very good for the crab of the sea,
But the wood of the crab is sauce
  for a drab who will not her husband obey.

Female sheet web spider Linyphia, *viewed in typical position below web and from above*

In Devon, young girls used to pick crab-apples from the hedgerows and lay them out in the shape of their suitors' initials. At daybreak on Old Michaelmas Day, 11 October, they would steal down and look at them: the initials in the best condition were those of their husband-to-be.

The folly of expecting good results from the most unreasonable acts is summed up by this old proverb: 'Plant the crab where you will, it will never bear pippins.'

The hedgerow and woodland crab-apples have a mixed parentage: some have undoubtedly descended from cultivated apples; others are native and truly wild. Most experts now agree that the tree consists of two sub-species: one the true crab-apple, the other, a native of south-east Europe and south-west Asia, from which the cultivated apples were developed. Our wild crab-apple is *Malus sylvestris* ssp. *sylvestris*, and the other is known as *Malus sylvestris* ssp. *mitis*. The two can usually be distinguished with reasonable certainty, as the cultivated apple and its wild relatives have leaves that are downy beneath, and with hairy leaf stalks and young twigs. The fruit is larger than the wild crab and often sweet. On the other hand, the leaves and twigs of our wild crab have very few hairs or are entirely hairless, and the fruit is always sour.

Rose hips are an important source of vitamin C; the syrup is a remedy for lethargy and has been claimed to be a preventative of the common cold. Rose-hip tea is an invigorating tonic, and in Tudor times they made mead with the scented berries of the briar rose, which was said to exhilarate the system. From the earliest times hips have been used in cooking tarts and pies.

It is known as the dog rose because of a quite false belief that the roots had the ability to cure rabies and the bites of mad dogs, so

110

the term 'dog' is not necessarily derogatory, although it often is when applied to wild or useless plants. Its local names include hedgy pedgies, puckies, choops, nippernails and soldiers. For folklore of the flower of the dog rose see page 61.

September is the month in which the only British shrub to produce edible nuts comes into season. The hedgerow hazel produces its familiar 'cobs' in clusters of one to four, and they may be destroyed by grey squirrels before they are even ripe. The greengrocers' 'filberts' are produced by another species of hazel that comes from south-east Europe and west Asia.

The hazel is a highly magical tree, rich in folklore and myth. Among many other things it is considered a symbol of happy marriages because its nuts are often united in pairs. Small twigs in the house protect it from lightning, and if you carry a double hazelnut in your pocket you will never suffer from toothache. Hazel rods are water-diviners, and also, for the ancient Greeks, an emblem of peace and reconciliation. With this and so much more pagan magic attached to it, the hazel was deemed worthy of a more sacred respectability, so in medieval Normandy it was christened 'Filbert' after St Phillibert whose feast day is on 22 August, when the hazel is supposed to come into nut.

Spiders

Few sights are more evocative of the onset of autumn than the early-morning festoons of dew-drenched gossamer draping the hedgerow and sparkling in the misty light. Careful examination will show that each individual sheet of gossamer is a spider's web, slung hammock-like between its supporting twigs and branches. They are made by money-spiders, the most numerous both in terms of total numbers and number of species of all British spiders. The family to which they belong, the Linyphiidae, contains more than 250 species, more than forty per cent of the total British spider fauna. An acre of meadow in autumn might well contain more than one and a half million of them. Most species are very small, about 2·5 mm long, and their accurate identification is definitely a job for the expert.

The commonest hedgerow species is probably *Linyphia triangularis*, which is larger than most – about 6 mm long. The female spins the web, which is slightly dome-shaped due to the upper 'scaffolding threads' pulling the sheet up in the middle. She can be found lurking upside down on the underside of the web patiently awaiting the arrival of her next meal. The threads of a money-spider's web are not sticky, and her prey is trapped when it blunders into the supporting threads and falls on to the sheet. Further struggles only serve to enmesh the insect more, and the spider then pulls it through the web from beneath. The hapless beast is wrapped in a silken purse and either consumed immediately or hung up on the edge of the web like poultry in a larder.

If the webs are examined carefully they may reveal the presence of a second spider often sitting towards the edge. This is the male who shares the web with the female for several weeks to enable mating to occur. For male spiders this is fraught with danger as he has to persuade the female that he is not just another candidate for the larder! Many spiders, including the money-spiders, have evolved a highly complicated ritual courtship which serves to disperse the female's aggression and coax her into being receptive. The male spider is easily recognised by the large inflated tips to his palps which are used for transporting the sperm during copulation. He is also relatively smaller than the female with a larger pair of jaws or 'chelicerae'. After mating, the female climbs down from the web and lays her egg-sac of about fifty eggs safe in the protection of the leaf litter in the hedge bottom, where they will remain until they hatch next spring.

Spiders are virtually unique in that, with the exception of man, almost no other group of animals has evolved the ability to construct a trap with which to catch its prey. This astonishing degree of evolutionary sophistication is matched only by a few species of caddis-fly, whose larvae construct what is in effect a trawl-net which is spread out over the floor of the streams in which they live.

Undoubtedly, the triumph of the spider's architectural genius is the orb-web, spun by the familiar garden spider. True to its name it can be found in almost any garden, however small, but it is equally common in the hedgerow. The spiders mature in August, and the webs, which are normally spun soon after dark, are a familiar sight at this time of the year. To watch the spinning of a web from start to finish is fascinating. The details of construction vary among species, so that an expert can identify the owner by the position and architecture of the web. The web consists of an outer polygonal frame of guy strands attached to the surrounding vegetation. From this a series of radial strands, meeting in the middle, form the support for the spiral of sticky silk that actually traps the flying insect prey. The first stage in the spinning of the web is the 'bridge-line'. The spider raises the tip of her abdomen and squeezes the silken thread out of her spinnerets to catch the breeze. The thread is wafted across a gap in the branches until it becomes attached to a twig on the other side. Firmly connecting a thread to the starting point and trailing this behind her, the spider then starts out across the temporary bridge and wraps it up in front of her as she goes. Having reached the safety of the other side, she firmly secures the end of the permanent bridge. The bridge-line may be strengthened by a further series of tight-rope excursions before the next radial stage is embarked upon; the spinning of the first three radial threads to establish the hub of the web. The sequence is shown above right. The spider fastens a loose separate thread to each end of the bridge-line, and from the middle of this she attaches a further strand and drops vertically to fix it to a suitable anchorage. The whole is then pulled taut. When the entire frame has been completed, the spider spins a temporary spiral from the centre outwards. She then retraces her steps back along the thread, but this time producing a special kind of silk coated with gum and destroying the temporary spiral as she goes.

Stages in the construction of a typical orb web

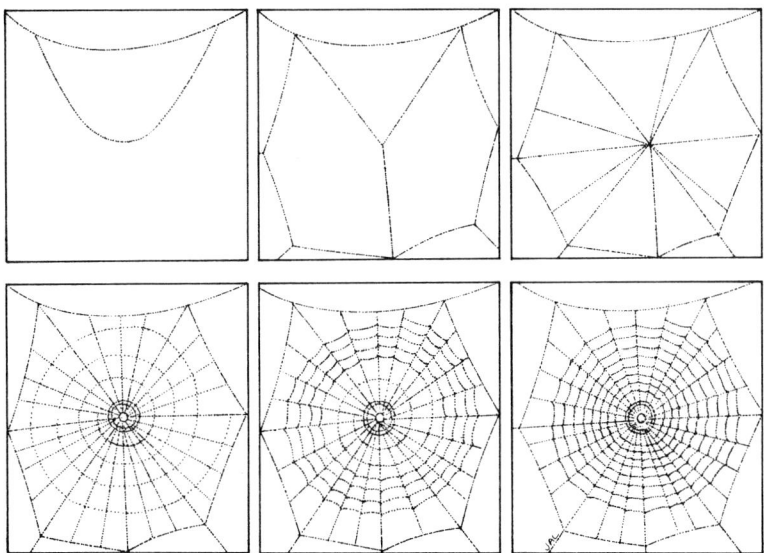

The completed orb-web consists of between twenty-five and thirty-five radial threads. Between the inner edge of the sticky spiral and the hub is a gap called the 'free zone', whilst the hub itself consists of a number of intermeshed threads, the 'strengthening spiral'.

The female spider rests head downwards on the hub, eight legs splayed to catch the welcome vibrations of a struggling captive – or else she may lie hidden in a nearby retreat in the foliage linked to the hub by a 'signal-thread'. Captured prey are bitten by the poison-fed jaws or chelicerae and then bound in a silken thread, bitten again and then carried off to the hub or the retreat.

At this time of the year the male orb-web spider may be seen searching for a mate. Cautiously he approaches the web, prudently trailing a well-anchored thread behind him. In all probability the female will repeatedly chase him off, causing him to swing well out of range on his escape line. Eventually his persistence is rewarded, and the female spider finally submits to his advances.

During September and October the female garden spiders leave their webs and go in search of a suitable cranny at the base of the hedge in which to lay their eggs. These are laid in masses of several hundred bound together by a covering of coarse yellow silk, to which may be added bits of leaf litter or other debris for increased protection and camouflage. Having survived the vicissitudes of winter, the eggs will hatch during a warm spell next May. The young spiders then set out on the hazardous journey that ends when they themselves produce the next generation of spiders.

The patient observer will sooner or later come across one or more of our twenty-odd species of harvest-spider, either clambering about the vegetation at the base of the hedge or actually among the leaf litter on the ground. These creatures bear a superficial resemblance to spiders, from which they can fairly easily be distinguished because the body is fused into one piece, instead of being divided into two distinct sections, the cephalothorax and the abdomen, as in the true spiders. Furthermore, the single pair of eyes is situated on

113

a raised hummock at the front end of the body, and they possess no silk glands so are unable to spin a web. Like the spiders, they are carnivorous, feeding on a range of small insects and other invertebrates, and they are most active at night. Several species are quite common in hedgerows, including *Mitopus morio* (in which the male has a distinct black mark on its back shaped like an hour glass), and the two excessively long-legged and small orange-bodied species of *Leiobunum*.

Bush crickets

An evening walk along any English hedgerow in late summer and autumn is almost certain to be enlivened by the chirping of bush-crickets. These insects are related to the more familiar grasshoppers, from which they differ in their extraordinarily long, thread-like antennae, much longer than the body, and the formidable looking cutlass-shaped ovipositor stuck on the end of the female's abdomen. This vicious-looking weapon is in fact quite harmless and is used to insert the eggs into the protection of some suitable crevice. Bush-crickets also differ from grasshoppers in that they are normally creatures of the dusk, a habit properly described as crepuscular (from the Latin *crepusculum*, twilight).

Another difference between grasshoppers and crickets is the way in which the 'song' is produced. Grasshoppers 'stridulate' by rubbing a row of teeth on the inner edge of their hind legs against the raised veins on the outer edge of their wing cases. Bush-crickets, on the other hand, chirp by rubbing their two wing cases together. The left forewing possesses a stiff tooth which is rubbed against the hind edge of the right, so creating the characteristic sound. It needs a keen ear to hear the bush-crickets chirping, as the noise is very high-pitched and people's sensitivity to the higher frequencies tends to decline with age.

If it takes a keen ear to hear them, it takes an even sharper eye to spot them. The most familiar hedgerow species is the dark bush-cricket (*Pholidoptera griseoaptera*) which is about 14–18 mm long, dark brown and with a brilliant lemon-yellow underside. The wing cases in the male are very short and only vestigial in the female. The best way to locate them is to creep up quietly to the spot from where the song seems to be coming. The chances are that the male (only the males chirp) will then stop for a while as he senses the disturbance. Stay motionless until he starts again, then peer carefully into the right part of the hedge and eventually he will be seen, probably sitting legs splayed on some broad leaf or else tantalisingly just hidden in the foliage. Greater patience still is needed to capture one, as at the slightest movement he will drop down deep into the vegetation – and as nettles are a favourite perch the odds are very much on his side!

The female lays her sickle-shaped eggs singly safe in the protection of crevices in bark or dead wood. Here they overwinter to hatch in the following May. The insect is fully mature by the end of July or the beginning of August, and in a fine autumn in

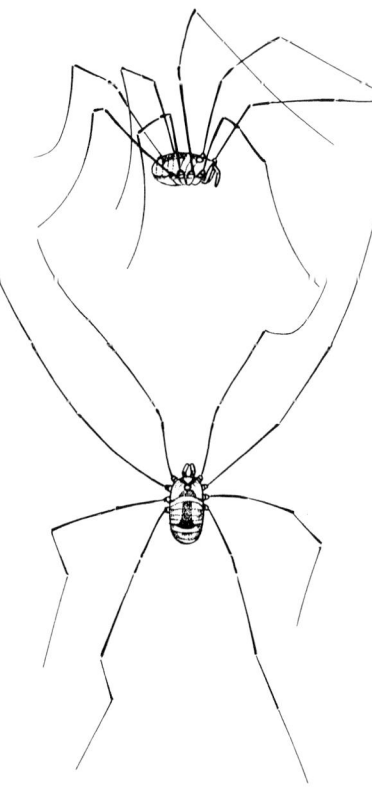

Harvest spider Mitopus morio

southern England they may be heard right up to the beginning of November, often singing all night. Like most of our bush-crickets, the dark bush-cricket is much commoner in southern England, and is not found at all in Scotland, Ireland and much of Wales.

Two more of our ten species of bush-cricket may also be encountered in hedges. The speckled bush-cricket is a bright emerald green, peppered with tiny dark-brown dots, and is especially common on brambles. It is often seen in gardens and occasionally finds its way indoors. The oak bush-cricket is also bright green in colour, but unlike the previous two is fully winged. As its name implies, it is our only true tree species, commonly found in oak but by no means confined to it. Furthermore, it is more strictly nocturnal than the other species and is unable to stridulate in the normal way. Instead, the male has evolved a most extraordinary way of making his presence known. The underside of the hind foot has a hardened pad, and this is drummed against the leaf that he is resting on in short rapid bursts.

Animals like the dark bush-cricket, which are now so much associated with hedgerows, again raise the question of their ecological origins. In this case, like the hedge-brown butterfly, there seems little doubt that we have an insect that originally would have been found on the shrubs around the edges of forest clearings, where indeed it is also to be found today. This is yet another piece of evidence that supports the view that in natural history terms hedgerows can be regarded as linear extensions of the woodland edge into the open countryside.

Pitfall traps

Because of their habits spiders and bush-crickets are relatively easy to observe, their webs or chirping eventually giving away their presence to the patient searcher. However, as we saw in July, much of the drama of hedgerow life is performed unseen among the foliage, and we have to employ special methods of collection, such as the beating-tray, if we are to become familiar with the individual actors. Huge as is the number of invertebrates grazing on the leaves of the trees and shrubs, in normal conditions they manage to consume only a relatively small percentage of the annual leaf production of the hedge, usually somewhere around 10 per cent.

The bulk of the foliage that is left uneaten falls to the ground in autumn, and is there set upon by the vast army of decomposer organisms that by the end of next year will have reduced the previous season's leaf fall to a mere amorphous organic residue and returned all the locked-up plant nutrients to the soil to fertilise the new season's growth. This process of decomposition is one of the most important in nature, and the colossal extent of the task that these decomposers accomplish can best be appreciated by considering what must be involved in disposing of a whole season's leaf fall in less than twelve months. The organisms responsible range in size from scavenging animals like woodlice and millipedes to the microscopic fungi and bacteria that are responsible for the

(Continued on page 118)   115

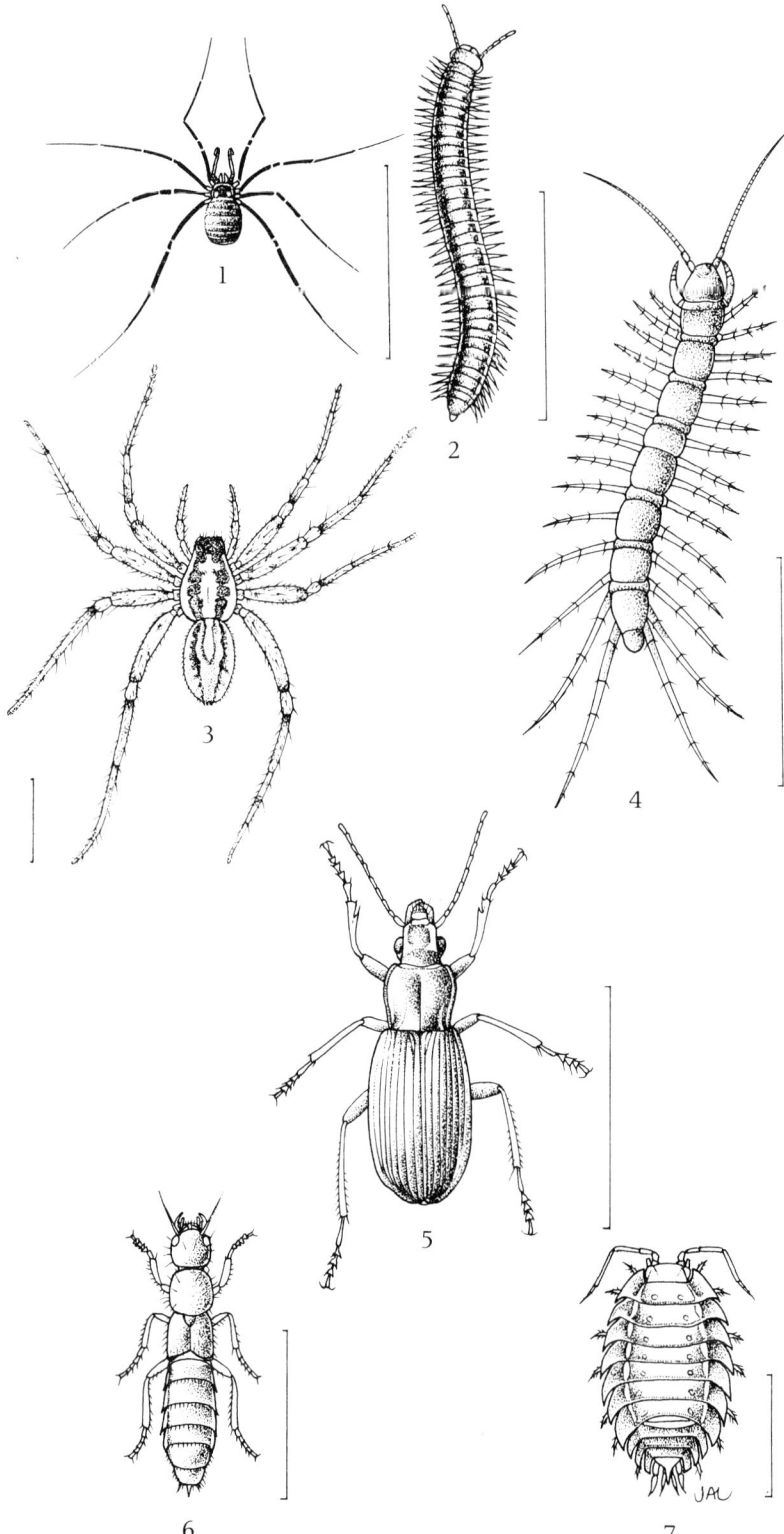

Selection of invertebrates from the hedge bottom

1 *Harvest spider*
  Nemastoma lugubre
2 *Millipede (Iulidae)*
3 *Wolf spider (Lycosidae)*
4 *Centipede* Lithobius
5 *Ground beetle (Carabidae)*
6 *Rove beetle (Staphylinidae)*
7 *Woodlouse* Oniscus asellus

Invertebrates of the hedge bottom

Ground beetles (Carabidae)
These are beautifully designed predators, with their long legs for speed, large eyes, long sensory antennae and large jaws. Nearly all are nocturnal, blackish in colour, and they range in size from 3 mm to 250 mm.

Rove beetles (Staphylinidae)
These are easily distinguished from ground beetles by their very short wing-cases which cover only the first two or three segments of the abdomen. Like the ground beetles they are voracious carnivores, and when threatened they will raise the tip of their abdomen over their backs like a scorpion – hence the common name 'cocktail beetle'.

Centipedes (Chilopoda)
As their common name implies, centipedes have a large number of legs, the British species having between fifteen and sixty pairs, one to each body segment. The head possesses a pair of powerful poison fangs, but none of our species is harmful to man.

Millipedes (Diplopoda)
The easiest way to tell a millipede from a centipede is by the number of its legs; a millipede has two pairs of legs to each body segment, whereas a centipede has one. The pill-millipede, which closely resembles a woodlouse, has seventeen pairs of legs, whereas a woodlouse has only seven.

Woodlice (Isopoda)
There are thirty-five different species of woodlouse in Britain, several of which are found in hedgerows. They range in size from the tiny *Trichoniscus* (4 mm) to the common pill-woodlouse, *Armadillidium*, that rolls itself up into a ball (18 mm). The commonest is *Oniscus asellus*, illustrated opposite, frequently called the slater, which is up to 15 mm long.

Spiders (Araneida)
The best adapted of the ground inhabiting spiders are the wolf-spiders (Lycosidae). As they spin no web, they rely on speed to catch their prey, hunting by day. The females carry their egg cocoons around with them attached to the tip of the abdomen. They are particularly active in warm sunny weather.

Harvest spiders (Opiliones)
Several species of harvest spiders inhabit the bottom of the hedge. The one illustrated, *Nemastoma lugubre*, is black in colour with a pair of white spots towards the front end of the body and has shorter legs than the vegetation inhabiting species described on page 113.

final breakdown processes. Some idea of the numbers of organisms involved can be gained from the fact there may be up to one hundred million bacteria and fungal colonies on a square centimetre of decaying leaf.

Inevitably, this huge litter-inhabiting community supports its own complex food-web of predators. Carnivorous beetles, centipedes, spiders and harvest spiders are in turn preyed upon by shrews and ground-feeding birds such as robins, blackbirds and song thrushes.

One of the easiest ways to collect the larger members of this dark and busy world is by means of pit-fall traps. The predatory members of the community hunt over the surface of the ground using their superior sight, speed and large jaws or mandibles. A simple pit-fall trap need be no more sophisticated than a jam jar sunk into the ground with the rim flush with the surface of the soil. Care must be taken to ensure that a 'moat' is not left around the outside of the rim or animals will fall into this before reaching the pit itself. A piece of flat wood covering the jar and supported on a couple of stones will help to prevent it from becoming filled with water in wet weather. As much of the activity of the hedge bottom goes on at night, several pit-fall traps set in the evening should yield an interesting catch by next morning.

As well as forming the basis of the whole complicated decomposer food-web, the leaf litter also gives protection from frost for the developing shoots of next spring's hedgerow flowers. Autumn-germinating annuals, such as goosegrass, can be found carpeting the decaying remains of last summer's plants whilst the thick layer of litter also provides cover for the hedgerow community of small mammals such as shrews, mice and voles.

The uses of hedgerow woods

The hedgerow harvest is not confined to its edible fruits, berries and leaves: there is also a harvest of timber which had many uses for the farmer and his wife in the past, although generally on a small scale, since timber for great ships and houses was obtained from woodland oak or elm. But many a farmer would plant in his hedge some of the large hard-wood trees, or else conserve existing trees that already stood there.

If the hedge was managed by coppicing, one of the most useful products to the farmer would have been hazel poles, cut every seven years or so. The lightweight, portable fencing called hurdles, vital in the heyday of sheep farming, was constructed almost entirely of panels of woven, cleft hazel poles, known as wattle work, and was made as soon as the wood had been cut while the poles were still pliant. Similarly, the wattle which formed the 'skeleton' of the 'daub and wattle' infill of mediaeval timber-framed houses was made from cleft hazel. It was also used for a whole range of other purposes about the farm, such as thatching spars or spics and for beanpoles and pea sticks. The waste from coppice of all kinds was used to fuel the bread ovens.

Old pollard preserved within a traditionally managed hedge

One of the most valuable of hedgerow trees is elm, deliberately planted in large numbers. Elm wood is specially favoured because of its resistance to decay and to splitting. Because it was so durable in water it was used for field drains and coffin boards, and because of its resistance to splitting the wheel-wright used it for the hub or nave of cart-wheels. A large hole had to be bored through the middle to take the axle, and an even number of slots cut around the edge to take the oak spokes. The rim of the cart wheel consisted of a number of individual curved segments, or felloes, of ash wood. Ash wood is both strong and resilient and is easily cleft and bent. It was, and still is, widely used for tool handles of all kinds, especially axes and hammers.

The wood of almost every hedgerow tree and shrub was put to a special use according to its particular properties. The hard wood of hornbeam was used for machine cogs, hawthorn for rake tines, the pale wood of holly and maple for turnery, spindle and dogwood for skewers, blackthorn for walking sticks and elder for pegs.

The most valuable of all woods is, of course, oak, and, like the elm, it was planted in hedgerows in large quantities, the trees often being managed by pollarding. Pollarding is essentially coppicing at

119

a height above the ground sufficient to prevent stock grazing the young shoots. The trunk of a pollarded tree is the 'bolling', and many old hedgerow oaks standing today have clearly been pollarded in the past.

Dead or unusable wood from hedgerow trees was a source of firewood, and even the resulting ashes have a use: they make subtle glazes for pottery, giving natural colours and interesting textures to the fired clay.

In some parts of the country special uses were made of particular hedgerow shrubs. For instance, the osier hedges of the Somerset levels, planted so that the field boundaries would remain clear during the winter floods, were used to make baskets of all kinds including the salmon traps or 'putchers', which are still set up in stacks in the Severn estuary. The putchers are shaped like long, conical baskets with the open ends set facing upstream to catch the salmon on the ebb tide. The putchers are being replaced by baskets of plastic-coated wire, but we were fortunate enough to watch one of the last of the traditional putcher-makers at work.

Clearly, there was far more involved in the traditional management of the farm hedge than the provision of a stock-proof fence.

Above
Salmon putchers in the Severn
Top
Pollarded willows marking a field boundary in flooded countryside

September Recipes

Cob-nut Canapés; Cob-nut Pancakes
Blackberry and Apple Dumplings
Bramble Muesli; Bramble Jelly
Melon with Blackberry Purée
Blackberry Rob with Soda and Ice
Bramble Drop-scones
Elderberry and Apple Crumble
Elderberry Wine
Crab-apple and Rose-hip Jelly

Cob-nut Canapés

110 g/4 oz cob-nuts, shelled
parsley or chives
2 cloves garlic
75 g/3 oz butter
salt

Chop the nuts finely and mix with the chopped herbs and crushed garlic. Work into the softened butter and season to taste. Spread on little squares of fried bread and serve with drinks as an appetiser.

Cob-nut Pancakes

(Makes 12 pancakes)

2 eggs
225 ml/8 fl oz
    water and milk, mixed
a pinch of salt
110 g/4 oz plain flour, sifted
10 g/$\frac{1}{2}$ oz castor sugar
25 g/1 oz melted butter

Liquidise the eggs, milk and water, and salt. Add the flour, sugar and butter, and liquidise for another minute. Leave to stand in a cool place, covered, for two hours. Thin out a little with milk if necessary.

50 g/2 oz cob-nuts
25 g/1 oz castor sugar

Chop the nuts very finely and add to the batter with the sugar. Make very thin pancakes and serve hot with cream and sugar.

Blackberry and Apple Dumplings

apples
blackberries
castor sugar
butter
sweet-crust pastry
egg

Core one apple per person and hollow out the hole a little. Roll some blackberries in castor sugar and fill the hole with them, pressing the berries down well. Put a knob of butter on the top and encase the apple in a square of sweet-crust pastry, rolled out thinly. Seal the edges carefully with water to make a parcel, and decorate it, if you like, with pastry leaves. Brush with beaten egg and cook at 375°F, 190°C or gas mark 5 for thirty five to forty minutes until golden. Sprinkle with castor sugar, and serve hot with cream or custard.

Bramble Muesli

Serves 2 to 3

1 tbs honey
a squeeze of lemon juice
110 g/4 oz crunchy muesli
275 ml/$\frac{1}{2}$ pint plain yoghurt
110 g/4 oz bramble
    jelly (see below)

Mix the honey and the lemon juice with the muesli. Add the yoghurt and mix together thoroughly. Finally, add the jelly and stir well. Serve immediately, for breakfast.

Bramble Jelly

blackberries
water
sugar

Simmer the blackberries with a little water until they are soft and mushy. Strain through a jelly bag overnight. To each pint of juice add $\frac{3}{4}$ lb/350 g sugar. Dissolve in a pan over a low heat and then boil to setting point (220°F/110°C). This will take from anything between twenty minutes to one hour, depending on the water-content of the fruit, and of course on how fast you are boiling it. Pour into hot, sterilised jars and seal when cold.

Melon with Blackberry Purée

blackberries
sugar and water
melon

Simmer blackberries in a little water with sugar to taste until softened. Liquidise all together, sweeten if necessary, then sieve to separate the pulp from the pips. This will give you a fairly thick purée which you can store in airtight jars, or freeze. Dice the melon and chill well. Mix with the blackberry purée.

You can make a thinner juice more suitable for a drink by straining the cooked fruit through a jelly bag overnight in order to extract all the liquid.

Blackberry Rob with Soda and Ice

8 cloves
1 stick cinnamon
110 g/4 oz brown sugar
570 ml/1 pint blackberry purée

Add the spices and the sugar to the blackberry purée, and simmer until thick and syrupy. Cool, then bottle and serve strained, either chilled with soda water and ice, or hot as a traditional remedy for colds.

Bramble Drop-scones

Makes 12

225 g/8 oz plain flour
½ tsp salt
2 tsps cream of tartar
1 tsp bicarbonate of soda
25 g/1 oz castor sugar
2 eggs
1 tbs golden syrup
225 ml/8 fl oz milk
175 g/6 oz ripe blackberries

Sift flour, salt and raising agents and add the sugar. Beat the eggs and add to the milk and the warmed syrup. Make a well in the middle of the flour and pour it in with the berries. Mix to a batter.

Put a tablespoon of the mixture on to a hot, well-greased frying pan or a griddle, and cook until golden on both sides and cooked through. Serve hot with butter.

Elderberry and Apple Crumble

Serves 6

350 g/12 oz elderberries
450 g/1 lb apples
110 g/4 oz sugar
110 g/4 oz butter
225 g/8 oz flour
110 g/4 oz light brown sugar
a pinch each of bicarbonate
    of soda and ground ginger

Pull the berries off their stalks and wash them. Core, peel and chop the apples. Mix them together with the berries and sugar and put into an ovenproof dish.

Rub the butter into the sieved flour, add the sugar, the bicarbonate of soda and the ginger. Press down lightly over the fruit with a fork and cook at 350°F, 180°C or gas mark 4 for thirty-five to forty minutes.

Elderberry Wine

(See notes on wine-making, page 104)

2 kg/4 lb elderberries
4·5 litres/1 gallon water
225 g/8 oz raisins
3 teaspoons citric acid
1 sachet wine yeast (Port type)
1·50 kg/3½ lb sugar

Strip the berries from their stalks, wash them in clean cold water and remove every trace of stalk and unripe berry. Place the berries in a suitable sterilised vessel, crush them and pour boiling water over them. Wash and chop the raisins and add to the vessel together with the acid. Cover and leave until cool.

Add the wine yeast and ferment on the pulp for three days, then strain out. Press and discard the fruit, stir in the sugar, pour the must into a sterilised fermentation jar and the excess into a large bottle, likewise sterilised. Fit an airlock into the jar and a plug of cotton wool to the bottle and leave them in a warm room until the wine is still – about three weeks. Move the wine to a cold place for a few days to help it to clear, then siphon it into a sterilised storage jar and another bottle. Bung tightly and keep for one year before bottling, then keep for a further six months before drinking. This is a strong, slightly sweet wine to serve after meals.

Crab-apple and Rose-hip Jelly

Using equal quantities of fruit, follow the method for bramble jelly. It makes a sharp jelly, good with game or on toast.

October

Here the industrious huswives wend their way
Pulling the brittle branches carefull down
And hawking loads of berrys to the town
Wi unpretending skill yet half divine
To press and make their eldern berry wine
That bottld up becomes a rousing charm
To kindle winters icy bosom warm
That wi its merry partner nut brown beer
Makes up the peasants christmass keeping cheer

John Clare, 'The Shepherd's Calendar'

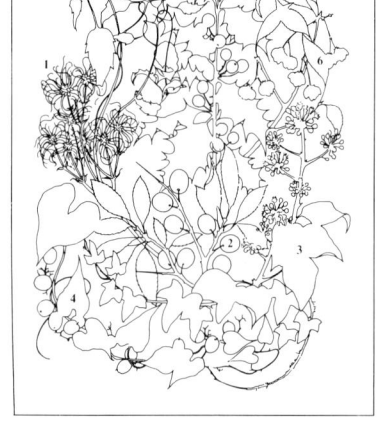

October is the month of harvest festivals, the month when the bounty of the land is being reaped and when people begin to prepare for winter. For Keats the 'season of mists and mellow fruitfulness', it often brings sunny days once the early morning mists have cleared, and there is a saying that October always has twenty-one fine days. The leaves start to change colour slowly as winter draws nearer: some years the autumn colours are vivid and violent, almost dazzling; other years they are soft, subdued and gentle as they fade from green to yellow, orange and red. The evenings draw in, and in between the last of the pale sunlit days come the first really cold ones. The wispy grey of old man's beard covers the quiet hedges, and autumn fungi thrive on the damp ground amongst the falling leaves. But even now there are whispers of spring: the leaves of next year's primroses peep through the ground, and cow parsley is shooting.

If the oak tree wears its leaves in October a hard winter may be expected: 'October with green leaves means a severe winter', so they say. The last day of October, Hallowe'en, is when the spirits of the dead are meant to appear. Traditionally in some parts of the country ceremonial fires were lit on Hallowe'en for the relief of souls in purgatory, and it is thought to be a favourite night for

1 *Traveller's-joy*
2 *Sloe (Blackthorn)*
3 *Ivy*
4 *Black Bryony*
5 *Holly*
6 *Spindle*

Above left
Bank vole
Above right
Common shrew
Below left
Wood mouse
Below right
Dormouse

witches to appear – hence the capers that are an integral part of Hallowe'en, from apple-bobbing to 'tricking and treating'.

Nuts and berries

It is common experience that some years are better than others for nuts and berries, if only from the annual pagan ritual of collecting holly to brighten the Christmas season. Interestingly enough, it seems that the years of the heaviest crops often coincide both between plants of the same species and also between different species of trees and shrubs. This is particularly marked with those that produce 'dry' fruits as opposed to those with 'fleshy' fruits or berries. A good 'mast' year for the oak, with a heavy acorn crop, quite often coincides with a good crop of hazel-nuts and hornbeam fruit. And hips, haws and holly berries are often particularly abundant or scarce in the same autumn. These differences are almost certainly due in large measure to variations in the spring and summer weather, different conditions favouring a 'nuts' year from that of a 'berries' year.

The autumn hedgerow harvest forms an important part of the diet of large numbers of different animals. This is not surprising when it is remembered that seeds are rich in the food reserves necessary to nourish the developing embryo and young seedling. Birds, particularly, feed voraciously on the hedgerow fruit crop, different species often showing marked preferences for different kinds of berry. Haws are particularly favoured by blackbirds, whereas the song thrush shows a preference for yew berries, if available. The mistle-thrush, on the other hand, has a particular liking for both holly and yew berries. Blackbirds will also feed avidly on crab-apples, as anyone who has left garden windfalls on the ground will know. Towards the middle of the month the first redwings will arrive, followed closely by the fieldfares from their breeding grounds in Scandinavia. Unlike the first three thrushes, both the redwing and the fieldfare move about the countryside in large, often mixed, loose flocks, and their distinctive calls are the real heralds of winter. Like the blackbird, both have a marked preference for haws, in addition to the insects and worms that form an important part of their diet while the weather remains mild enough.

It is not only the thrushes that are attracted to the deep-red clusters of hawthorn berries. Flocks of greenfinches are also a familiar sight at this time of year, the bright yellow patches at the base of the tail and on the front edge of the wing showing up conspicuously as they flit among the higher branches. However, whereas the thrushes were after the soft, fleshy part of the berry, the greenfinches' interest is the nutritious kernel which lies inside the stone. Their specially designed, broad, heavy beak is able to crack the stone in a matter of seconds, a feat that most people are quite unable to achieve without a subsequent visit to the dentist! Once the haw crop has been exhausted, usually by early in the New Year, the greenfinches can be expected to make an appearance on the tits' peanut bag in the garden.

127

Mammals

Birds are not the only ones to feast on the autumn berry crop. A careful search beneath the hawthorn in late autumn will reveal many of the small stones from the inside of the berry with a neat round hole gnawed in it, indicating the attentions of wood mice or bank voles.

The presence of mammals in the hedge is more easily inferred from the indirect evidence of their activities, such as food remains, footprints and droppings, than by direct observation. The most numerous are the mice, voles and shrews. Although superficially similar and often referred to as 'mice', shrews are not at all closely related to the true mice and voles. They are primarily carnivorous, feeding almost wholly on a wide range of insects and other invertebrates and as such are closely related to the mole and hedgehog. Mice and voles are typical rodents, with a predominantly vegetable diet supplemented by animal food.

The five commonest species of small mammal to be found in the hedgerow over most parts of Britain are the wood mouse, bank vole, field vole, common shrew and pygmy shrew. The characteristic external appearances are shown opposite, but the main differences between the rodents and the shrews lie in the teeth and the fore-feet. Shrews have a continuous row of teeth, but in the mice and voles there is a gap between the front cutting teeth (incisors) and the chewing or cheek teeth (molars), typical of most vegetarian mammals. In addition, shrews have five toes to their fore-paws, and mice and voles have only four. Their small size – the pygmy shrew is our smallest mammal – together with the long, tapering snout and small eyes, should serve to distinguish the shrews from the mice and voles on superficial appearance.

Wood mice are at their most numerous in the early autumn. They are agile climbers and are particularly fond of rose hips, which they collect by gnawing the stalk diagonally across just below the base of the fruit. They are also fond of the kernel of the hawthorn berries, which they extract by gnawing a circular hole in the stone. Hazel-nuts are a favourite food of both wood mice and bank voles, and it is sometimes possible to tell from the empty shells which of the two it is that has been feeding on them. Both hold the nut with the base pressed to the ground with their fore-paws, but the wood mouse leaves a ring of tooth-marks around the outside rim of the hole whereas the vole tends to leave a clean outside edge to the hole as shown on the right. Wood mice are essentially nocturnal animals, their prominent black eyes and large ears being well adapted to such an existence. As their name implies, they are characteristically a woodland animal, but they are also common in hedgerows, fields and gardens and are happy in habitats with less ground cover than the bank vole. They vary their diet with buds, seedlings, various insects and also snails, which they extract by biting through the shell.

In southern England and parts of Wales a second mouse, the yellow-necked mouse, may occur in the same sort of habitats as the

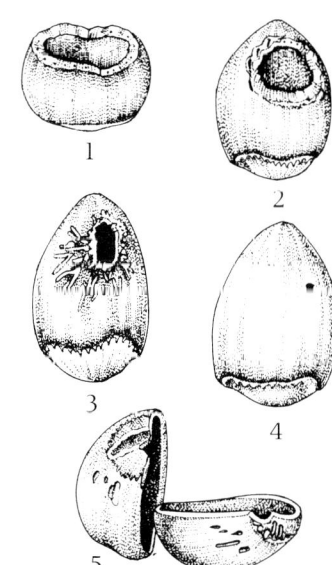

Hazel nuts attacked by

1 *Bank vole*
2 *Wood mouse*
3 *Bird*
4 *Weevil*
5 *Squirrel*

Small mammals of the hedgerow

1 *Wood mouse*
2 *Bank vole*
3 *Field vole*
4 *Pygmy shrew*
5 *Common shrew*

129

wood mouse. It is larger, with a longer tail, and it has a conspicuous yellow band stretching across its chest between its front legs. Like wood mice, they often come into buildings, such as barns and apple lofts, during the winter.

The bank vole is very much an animal of the woodland edge and hedgerow, requiring a great deal more ground cover than the wood mouse, which is hardly surprising as it is far more diurnal in its habits. The rounded muzzle, small ears and short tail at once distinguish the voles from the mice, and the bank vole has a rich chestnut-coloured coat and a tail which is about half the length of the body. It is more vegetarian than the wood mouse and prefers the flesh of the haws to the kernel.

The short-tailed vole is less often found in the hedgerow, being predominantly an animal of rough grassland, but it does resort to the hedge when the surrounding fields are disturbed by cutting or ploughing. It feeds almost wholly on grass, and its numbers can fluctuate enormously from year to year. It is browner than the bank vole, has a shorter tail, and the small ears are almost completely hidden in its fur.

Both mice and voles form an important part of the food of a number of predatory birds and mammals along the hedgerow, particularly tawny owls and weasels. Shrews, on the other hand, appear distasteful to most predators, only owls including a significant proportion of them in their diet. Shrews are active throughout almost the whole of the twenty-four-hour period and consume about three-quarters of their own body weight each day – more when the female is suckling young. Their food consists almost entirely of the insects, worms and woodlice that they find with their long, mobile snouts among the leaf litter at the bottom of the hedge. The common shrew is more abundant than the pygmy but less widespread. The high-pitched squeaking that one so often hears in the hedge undergrowth, though rarely seeing the perpetrator, is almost certainly made by shrews. They are solitary beasts, and chance encounters between individuals can be noisy affairs!

The animal that is eponymously associated with the hedgerow is of course the hedgehog; in some country areas it is rather rudely known as the 'urchin'. They are probably commoner than is often thought, as they are almost entirely nocturnal. An extraordinary medley of snorts and grunts, together with a great deal of rustling, gives away the presence of a foraging hedgehog in the dark. Although hedgehogs hibernate in a nest of dry leaves and moss in the bottom of the hedge, they are occasionally active in mild weather throughout the winter. Surprisingly, hedgehogs can both swim and climb quite well.

In spite of their spiny armour, hedgehogs are preyed upon by foxes and badgers, both of which often take up residence in the hedgebank, especially in sandy or chalky areas. We found quite a large badger set at the bottom end of the Sussex hedge, conveniently close to the stream. Over the years the animals had dug a complex of burrows with several entrances, some of which emerged some way into the field. It was easy to follow the well-marked paths that they had beaten into the wood and down to the stream. The typical, broad, five-toed footprints were clear in the bare sand outside the set.

It is quite a simple matter to make a collection of plaster casts of animal footprints. All that is needed are some strips of card, paperclips, plaster of Paris, water and a container in which to mix the plaster. Cut the card into a strip about 8 cm wide, and fix it in a circle around a suitable footprint with the paper-clips. Mix the plaster by first pouring the water into the container, and add the plaster until it reaches the surface of the water. Stir well and pour into the mould. Leave to harden for about half an hour, remove the card and clean off the dirt from the print. If desired, the pad, toe and claw prints can be blacked over – and always be sure to label the reverse of the print with the date and location.

Undoubtedly the most delightful, but sadly now the scarcest of our hedgerow mammals, is the dormouse. Its foxy-red colour and long, bushy tail, together with its arboreal habits, all conspire to give it a quite special charm. Its wholly nocturnal habits make it an even more difficult animal to get to know. It feeds largely on nuts, especially hazel-nuts, and the young shoots of shrubs, but its presence is most often revealed by the stems of honeysuckle from which it strips the bark to construct its dome-shaped nests.

Other mammals that inhabit the hedge and hedgebank include the brown rat, especially during the summer, and the rabbit. Rabbits form a major part of the diet of both foxes and stoats, which the patient observer may be fortunate enough to see hunting along the bottom of the hedge in search of unwary prey. Thus a good length of old hedge in rural areas can be expected to harbour at least a dozen different species of mammal, which, if you exclude the bats, is more than a quarter of the total British mammal fauna.

October plants

Spindle is a tree of good repute, according to the translation of its generic name, *Euonymus*; but it is more commonly known as an unlucky tree. Perhaps it got its name, like the Cape of Good Hope did, with a view to propitiation. Its local names describe its many uses, for spindle is a hard white wood which has been used in the past to make spindles, skewers and pegs – hence spindlewood, skewerwood and pegwood. The burnt twigs make excellent char-coal drawing sticks, and the berries, dried and powdered, are used to kill ticks, lice and other parasites, particularly in the hair.

It is a poisonous plant, a violent emetic: it has been used as a purgative in folk medicine, but its usefulness lies in the external treatment of scabies and related conditions. Sheep and goats have been poisoned by eating spindle – hence another local name gatter-tree, meaning 'goat-tree'. Spindle's four-lobed, orange-pink berries are its crowning glory: in the words of Tennyson, 'fruit which in our winter woodland looks like a flower'.

The festoons of sealing-wax red berries of the black bryony remain the most colourful part of the hedge for much of the winter in England and Wales. The berries escape the attentions of hungry birds and mammals as they contain a powerful poison, saponin. Most berries are attractive to animals, and this helps in seed dispersal. In the case of the black bryony, on the other hand, it seems that the function of the berry is to protect the seeds from desiccation, because the seeds will germinate only after a long period of wet weather. Like the holly, black bryony is dioecious (that is, it has separate male and female plants). It is classified as a poisonous plant, since the root is violently purgative and the red berries are so powerfully emetic that they can be fatal to children.

Black bryony is the black vine of ancient medicine as opposed to white bryony, the white vine, but the two are unrelated except in name. Black bryony has an exotic ancestry, being the only wild representative of the yam family native to Britain. It possesses a large starchy root not unlike that of its cousin, the vegetable yam, which is an important food plant in Africa, Asia and the West Indies. Its common names include chilblain berry, poison berry and snake's food, and a poultice made from the berries was laid on chilblains to soothe and cool them, the berries being stored in alcohol throughout the winter for this purpose. They also remove freckles, sunburn, black eyes and skin blemishes. The root has a use in homoeopathic preparations for rheumatism and sunstroke.

Above far left
Stoat
Above left
Badger
Below left
Hedgehog

133

It was John Gerard who coined the name 'traveller's joy' for the well-known climber so decorative in our autumn hedges, because of its 'decking and adorning waies and hedges were people travell'. He saw it in every hedgerow from Gravesend to Canterbury, 'making a goodly shadow' beneath which travellers could rest: 'thereupon I have named it Traveller's Joy'. He also called it 'virgin's bower', considering it a suitable plant to form an arbour in which young girls could sit, protected from summer sunshine and from passers-by. In the Victorian 'language of flowers' it stood for rest and safety. Its other very common name is old man's beard, partly because of its feathery tufts, and partly because this Old Man, alias the Devil, can twist around trees and shrubs and choke them to death. It is also called 'boys' bacca' because you can smoke cigar lengths of the stems as they are hollow and do not easily burn.

Its Latin name is *Clematis vitalba*. '*Clematis*' comes from a Greek word meaning 'vine-shoot', from the resemblance of its branches. '*Vitalba*' comes from a Latin word for vine, and '*alba*', white, from its fruits.

The Holly and the Ivy
When they are both full grown
Of all the trees that are in the wood
The Holly bears the Crown.

The holly tree with its red berries is a universal favourite, perhaps because it is so closely connected with Christmas, as its local names 'prickly Christmas', 'Christmas' and 'Christ's thorn' imply. Mediaeval monks called it the Holy Tree, a name which stuck until the end of the seventeenth century, and through its Christmas associations it has become a symbol of eternal life. We have long decorated our houses at Christmas with branches of holly, and folklore has it that elves and fairies join in the celebrations if they can shelter in these holly branches, in return for which they will protect the inhabitants from the antics of the house goblin. It is unlucky to bring holly into the house before Christmas Eve and to take it down before Twelfth Night: the traditional date for stripping the house of its decorations is Candlemas Eve, 1 February.

As far back as the days of ancient Greece it was thought that a holly tree planted near a house would protect it from lightning and keep away evil spells and enchantments. Since the holly leaf is prickly it drives away enemies, and witches are said to detest it because it is holy. In the dead of the year its berries are a protection from evil, red being a potent magic colour. It is a tree to be deeply respected:

Who so ever against holly do sing
He may wepe and handes wring.

A smooth-leaved holly leaf is placed under the pillow for the divination of dreams, and in the 'language of flowers' holly stands for foresight and good wishes. Folklore has a barbaric-sounding cure for chilblains, which prescribes thrashing them with holly twigs. There is also an old wives' tale that if children drink milk out of a holly cup they will be cured of whooping cough. The heart wood of holly is a hard, white, fine-grained wood with an ivory sheen and is a favourite with turners and engravers. Traditionally chessmen were carved out of holly.

The holly has given its name to many towns and villages: 'holm' is an old name for holly and features in Holmbury, Holmswood and Holmsdale, for example.

It is sometimes a cause of puzzlement or even perhaps irritation that one's most accessible holly bush never produces berries. Holly, like the various species of sallow, has separate male and female trees, so male plants can never produce berries. It is generally assumed that the prickly leaves have evolved to discourage grazing animals, and certainly the leaves at the top of the older bushes are hardly prickly at all.

Ivy is perversely in full flower during October. The small, five-petalled, yellow-green flowers are arranged in tight umbels and are very attractive to the early autumn flies and wasps. Ivy leaves are only ivy-shaped on the younger parts of the plant. Leaves on the older shoots are unlobed, and it is also those shoots that produce the flowers.

It is a magical plant and used to be as much a feature of Christmas decorations as holly, a sadly-lapsed tradition since the leaves and berries possess a special beauty. It also shares some of holly's protective properties. It is a kindly plant, supportive and feminine in its clinging habit, and it provides nesting cover for birds without damaging the tree or wall up which it grows. It is not a parasite since it roots extensively in the ground and has no need to penetrate deeply the surface to which it clings. Ivy grown up the wall of a house will protect the inhabitants from witches, but if it withers it is an omen of disaster.

The ivy, which protects and embraces, has become an emblem of confiding love and friendship, and in ancient Greece the altar of Hymen was encircled with ivy, and a branch of it was presented to a newly-wed couple as a symbol of their indissoluble knot. It represents love, constancy and dependence and is, along with holly, a symbol of immortality. In the 'language of flowers' it stands for fidelity and friendship and has been used as a love-charm: a country girl would place a twig of ivy leaves in her pocket or bag, and go out walking with the conviction that the first man who spoke to her would become her husband. Also, ivy was sacred to Bacchus, the god of wine, life and vitality, who was crowned with a wreath of ivy and vine leaves, because as a boy his mother hid him under an ivy bush. Ivy-berry vinegar was a popular remedy in the Great Plague of 1665; and an infusion of the leaves relieves sore eyes. A decoction from the berries is said to be a remedy for rheumatism, but the berries are poisonous if taken in large doses. Ivy leaves, boiled and mashed until the water is dark, makes a rinse for revitalising black silk.

Hedge laying

Most hedge management is traditionally carried out during the autumn and winter, although it is advisable to avoid hedging in periods of heavy frost to avoid damage to the bushes. It should always be remembered that the main purpose of good hedge management is to provide an effective, live, stock-proof barrier.

The most usual and widespread form of early management was probably coppicing, where the shrubs were cut back close to the ground to form a 'stool' every twelve to fifteen years. This provided the farmer with a regular supply of wood products, such as fencing, hurdles, thatching spars and firewood. The management of the hedge would have been closely integrated with the general management of the farm, so that in the early years after coppicing the surrounding fields were put to arable and not returned to pasture until the hedge had grown up sufficiently to produce a stock-proof barrier once again.

The traditional cut and laid hedge was probably not widely established until the eighteenth century and the time of the Parliamentary enclosures when farm leases began to include specific clauses on hedge management.

Today traditional hedge-laying is largely restricted to the Midlands, the South-west and Wales, although examples can still be found in other areas like the Weald. It is a craft which requires a high degree of skill – and sadly, with the advent of mechanical cutting, few now practise it. Each part of the country has its own distinctive and recognisable style of management, based primarily on whether the hedge is designed to be a sheep fence or a cattle fence. A sheep hedge needs to be thick and impenetrable to prevent the animals scrabbling through the bottom, whereas a bullock hedge has to be strong enough to withstand a good deal of heavy shoving and leaning. Bullock hedges are commoner in the Midlands, while the Welsh and border hedges are mostly sheep fences.

In order to encourage and keep alive the craft, there are now well over thirty hedge-laying competitions held each year, often sponsored by the local hunt or Young Farmers Clubs. We were fortunate enough to visit the National Hedge-laying Competition organised by the National Hedge-laying Society and held at Orlham Farm near Ledbury in Herefordshire.

The competition is divided into two main classes – the Midland and the Welsh Border styles. Each competitor has to lay ten yards of hedge in five hours. The main tool of the hedge-layer is the billhook, which reflects as many regional patterns as the hedges themselves. In addition axes, slashers and, more recently, chain saws, all form part of the standard equipment.

The familiar laid hedge is the Midland bullock hedge. It consists of stakes, pleachers and hethers or binders. The most surprising feature of this kind of hedge-laying is the amount of the hedge that is cut out before the selected shoots are laid. There seems to be hardly any hedge left to work on. The stems of the remaining bushes are first partly cut through close to the base and

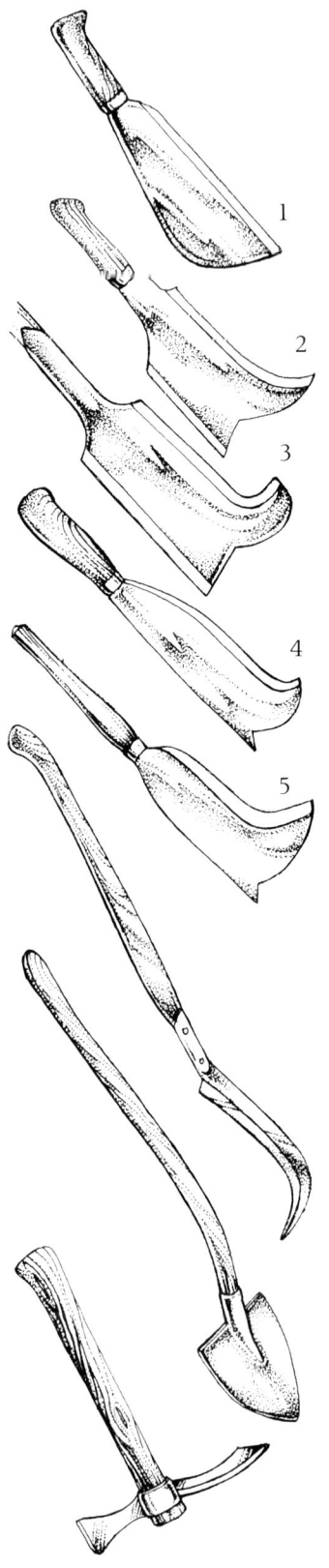

1

2

3

4

5

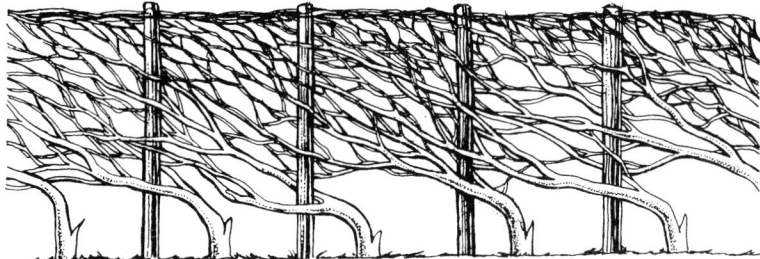

then bent over or laid so that they lie flat or diagonally upwards with the lowest stem as close to the ground as possible. The 'brush' side of the stem or 'pleacher' is laid so that it faces the field side of the hedge. The hedge is then staked with ash or hazel poles set about eighteen inches apart in line along the ditch side. Hethers or binders, long pliable withies of hazel, are then pleated between the stakes along the top of the hedge to provide a finish and to keep the pleachers in place, and finally the stakes are topped off to an even height of about four feet by an upward, diagonal cut of the bill-hook. The job is completed by trimming up the brush side, cleaning up the rubbish and clearing out the ditch.

The Welsh hedge varies enormously in detail, each county claiming its own characteristic tradition. However, because they are primarily sheep fences they are always dense but not necessarily very tall. In comparison to the Midlands hedge. they are often 'double brushed' – that is, the pleachers are laid in from both sides of the hedge so that the bushy ends project alternately on one side and then the other. Various amounts of dead wood can be added to pack out the hedge until it thickens out by natural growth. In some areas neither stakes nor hethers are used, but Welsh hedges are always beautifully finished and close-trimmed.

Further details of style and management can be found in *Hedging*, the comprehensive manual produced by the British Trust for Conservation Volunteers (see page 153).

Other parts of the country have developed different hedging traditions, such as the characteristic turf bank and hedge of Devon and Cornwall, requiring its own specialised tools, the mattock and Devon shovel.

The competition hedge at Orlham Farm had last been cut by the farmer about thirty years previously. This is a reasonable period between successive laying in parts where the hedge is fairly slow-

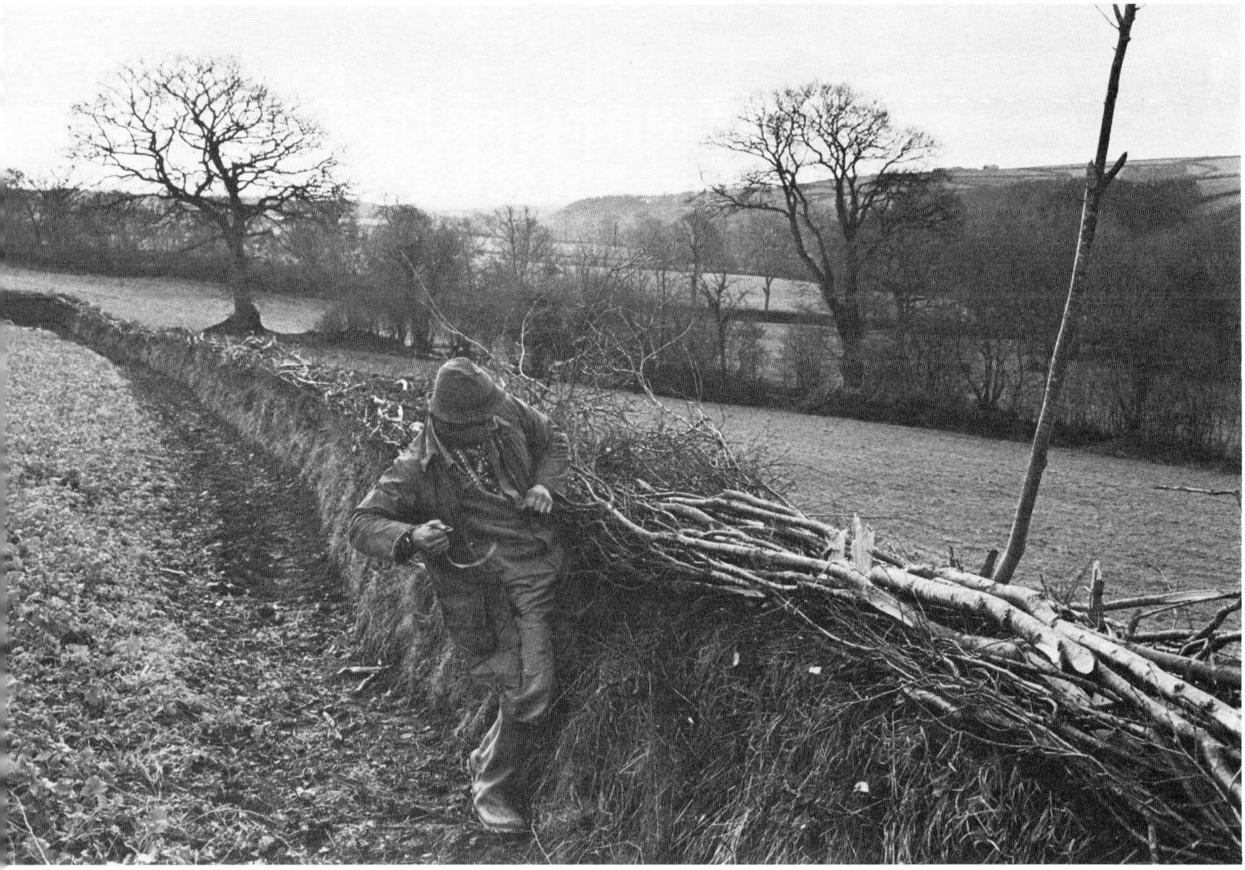

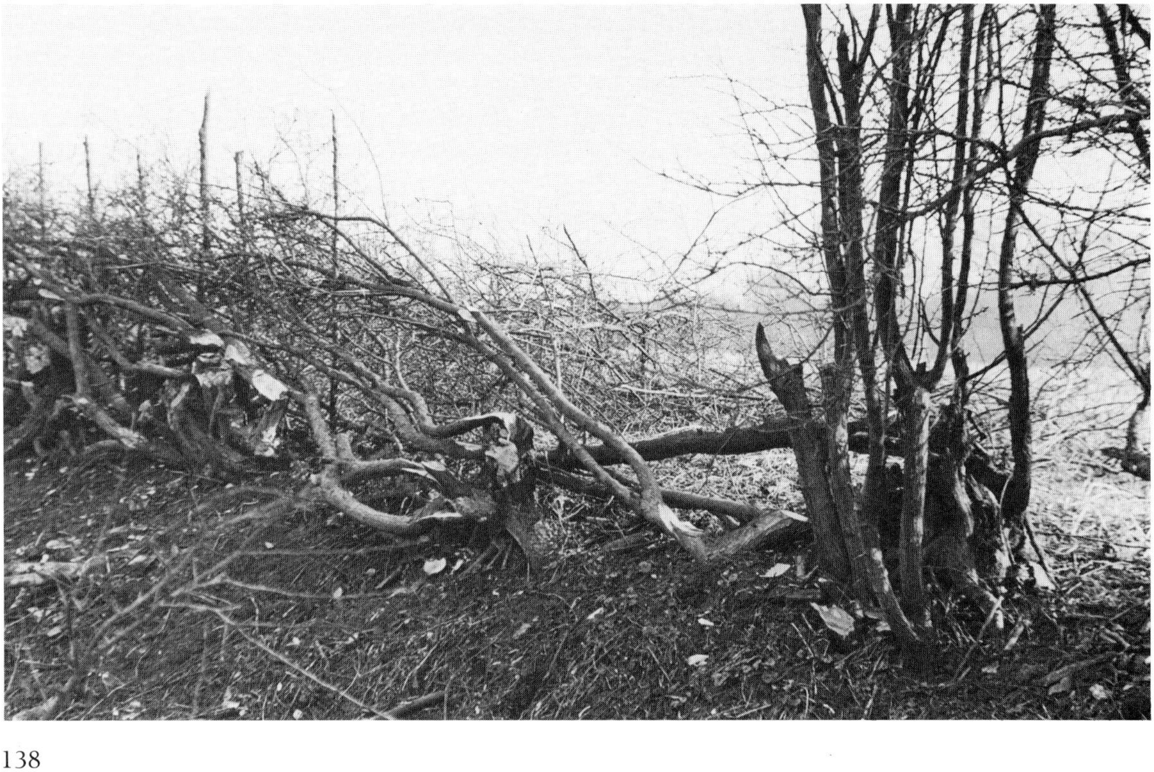

Above
*Mechanical hedge trimming
using a tractor-mounted flail*
Above left
*Laying a west country hedge
and bank*
Below left
*Gappy hedge being restored
by traditional laying*

growing and is trimmed regularly. In other areas hedges may have to be laid every fifteen to twenty years or so, if they are not to become 'gappy' at the base.

Hedge-laying and -trimming is obviously a skilled and time-consuming job, so that it is not surprising that most farmers now cut by machine, using either a tractor-mounted shape-saw or flail. Inexpert use of badly maintained machinery can, and often does, result in a butchered wreck of a hedge with unsightly flayed and ragged branch ends. However, it is possible with care and practice to trim a hedge cleanly and efficiently with a mechanical flail, as we saw on a fruit-farm in Gloucestershire. It is even possible to trim round standard trees, but the biggest problem is the wire stays of telegraph poles. With mechanical cutting being between ten and twenty times cheaper than hand trimming and anything up to seventy times cheaper than cutting and laying, farmers are hardly encouraged to rely on traditional techniques in the present economic climate. Furthermore, ugly as the results of careless mechanical cutting may be, there is little evidence so far of permanent damage to the hedge unless it is cut too far back on the face. This encourages the growth of hedgerow 'weeds' such as bindweed and black bryony, which may hinder the regrowth of the young woody shoots. With time the hedge may also become thin and gappy at the base.

October Treats

Sloe Gin, Sherry and Brandy
Chocolate Sloes
Bullace Vodka
Blackberry Vodka Cocktail
Blackberry Ice Cream
Blackberry and Elderberry Ice Cream

With the prospect of cold days and dark evenings ahead, now is the time to gather the last of the fruits of the hedgerows to make some cheering drinks and treats for Christmas. The companionship of a glass of sloe gin or of bullace vodka is a welcome friend in those long and seemingly interminable winter evenings.

Sloe Gin

225 g/8 oz sloes
50 g/2 oz sugar
6 unpeeled almonds
570 ml/1 pint gin

Pick the sloes after the first frost, and wash them. Remove all the stems and leaves, and prick the berries with a fork. Drop them into a bottle until they come about one-third of the way up. Pour in the sugar through a funnel, add the lightly bruised almonds and then pour in the gin. Give the bottle a good shake and give it further daily shakings until the sugar has dissolved. Leave for three months, and then strain off into another bottle and store for one year before drinking.

It is a shame to throw away all those gin-soaked sloes when you have strained off the gin, so below are some ideas for using them.

Sloe Sherry

175 g/6 oz gin-soaked sloes
75 g/3 oz sugar
450 ml/16 fl oz dry sherry

Put the sloes into a bottle and add the sugar. Pour the sherry over the fruit and shake well daily until the sugar has dissolved. Leave to stand for about a month, and it is ready to drink.

Sloe Brandy

Make this in exactly the same way as the sloe sherry above, but use brandy instead: it is delicious.

Chocolate Sloes

175 g/6 oz gin-soaked sloes
225 g/8 oz plain chocolate
rice paper

You can make your own home-made liqueur chocolates with the left-over gin-soaked sloes: just chop the flesh of the fruit and add it to the melted chocolate. Let it set on rice paper and then cut it into squares. Store in an airtight tin.

As a variation, try adding chopped cob-nuts to the melted chocolate as well as the sloes: it makes an extra-special treat.

Bullace Vodka

Pick your bullaces, either golden or black, and wash them. Make the bullace vodka in exactly the same way as sloe gin (see page 140), omitting the almonds, and substituting vodka for gin. It is best drunk within six months.

You can also make *Blackberry vodka* along the same lines, straining it off the fruit after one month. It turns a beautiful clear deep red.

Blackberry Vodka Cocktail

Something to warm up with on a cold day! The blackberry juice (see page 122) freezes very well, and it is worth making enough in September to keep you supplied through the winter. Add twice the quantity of blackberry purée to a measure of vodka and you have a kind of sister to a Bloody Mary.

Two more recipes to go into the freezer for the winter ahead when the hedgerows are no longer abounding with fruit:

Blackberry Ice Cream

275 ml/½ pint double cream
25 g/1 oz icing sugar
275 ml/½ pint blackberry
    purée (see page 122)
2 egg whites

Whip the cream until it is thick, and fold in the sieved sugar. Then fold in the blackberry juice. Whisk the egg whites until they are stiff and fold them into the cream mixture. Freeze.

This mixture is also delicious before it is frozen, served chilled with tiny meringues.

Blackberry and Elderberry Ice Cream

350 g/¾ lb each blackberries
    and elderberries
275 ml/½ pint double cream
40 g/1½ oz icing sugar
2 egg whites

Cook the fruits together with a little water and enough sugar to sweeten. Liquidise them when they are soft, and then sieve to separate the pips out from the pulp. Whip the cream until it is thick, fold in the sieved sugar and then the fruit purée. Whisk the egg whites until they are stiff, fold them in and freeze.

The Way Ahead

The countryside in winter looks misleadingly lifeless, and nowhere is this more true than in the hedgerow. The leafless skeletons of the trees and shrubs standing stark against the grey sky heighten this illusion. Yet next February's hazel catkins are already about 2 cm long, and at the base of the hedge many spring-time plants are encouraging thoughts of longer days. Goosegrass seedlings are germinating around the decaying remains of last summer's nettle beds, the basal leaves of cow parsley and other hedgerow biennials are already quite large and the young shoots of primroses are beginning to push their way up through the soil. Activity in the hedge bottom is still pretty frantic. Very few of our small mammals go into true hibernation, with the exception of the now scarce dormouse and possibly the hedgehog, and the extent of their comings and goings is easy to see during a walk after a fall of snow. Large flocks of birds patrol the hedgerow in search of berries, seeds and the wintering stages of insects.

Winter is actually a good time to start a study of a local hedge. Basic surveying is easier when the vegetation has died down, and there is less chance of disturbance to the wildlife. Of course it is simpler to record the trees and shrubs when they are in leaf, but with a little practice it is soon possible to become proficient at identification from the winter twigs. The illustrated key on pages 149 and 150 will help. It should be used in exactly the same way as the key to grasses in the June Chapter. Be sure to use the text and the illustrations in conjunction, as you are likely to come up with a wrong answer if you try to identify a twig using the illustrations alone. Particular attention should be paid to the way in which the buds are arranged on the shoot – whether they occur in opposite pairs on either side of the stem or whether they are arranged in some other way. The colour of the buds and the bark is also important. Some twigs, like the hazel, are distinctly hairy, although in others the bark is smooth and glossy, often with numerous small, distinct, pale-coloured warts called lenticels. The shape and size of the scar left by last summer's leaf may also be characteristic.

Autumn and winter are also the times of year when most hedge maintenance is carried out, both because there is more time available from other farm activities and also because the hedge is less

Above right
Spindle
Above far right
Sloes (Blackthorn)
Below left
Rose hips
Below far left
Haws

142

likely to be damaged while it is dormant, except in periods of heavy frost. However, in many parts of the country the traditional hedge landscape has been transformed over the last thirty years. In order to understand the reasons for this it is necessary to remind ourselves of the traditional purpose of farm hedges. Primarily, they functioned as stock-proof fences, their regional variations reflecting differences in the dominant type of farm enterprise; sheep fences in Wales and the Borders, bullock fences in the Midlands and so on. Secondly, they doubled as visible boundaries between individual farms, estates or parishes. Hedges also act as shelter for both stock and crops and also to help combat the problem of soil erosion in exposed situations. Lastly, as we saw in September, the hedge was traditionally managed with great care in order to provide a continuing supply of underwood for a whole range of essential purposes. The significant point is that hedges were established and maintained because they performed an important agricultural function, and any landscape or wild-life benefit was largely incidental. However, the revolution in British agriculture that really got under way in the early 1950s profoundly altered all that. Increasing mechanisation and a shift in emphasis to intensive arable farming in many parts of lowland Britain had far-reaching implications for the hedgerow. Small enclosed fields were no longer compatible with large, modern combine-harvesters and intensive cereal production. In order to produce one extra acre of land for cultivation it is necessary to remove 2,420 yards of a six-foot-wide hedge – or to produce an extra hectare, 5,000 metres of a two-metre-wide hedge.

A natural accompaniment to the increasing mechanisation of the farm is the progressive reduction in the labour force. Skilled workers are now simply not available on many farms during the slack period to carry out the necessary hedge maintenance. Traditionally, hedges were trimmed back in the early spring before the sap began to rise and again in the autumn. Laying was only necessary when the hedge began to get 'gappy' at the base, and this might have been at anything between ten and thirty years, depending on local conditions. As a result, as hedges deteriorated they tended to be replaced by post-and-wire fences which need less maintenance, but on the other hand don't last anything like as long. A hedge, if properly cared for, will last indefinitely.

The importance of a hedge as shelter for stock naturally disappears with a conversion from pasture to arable. At the same time, however, the vulnerability of the land to soil erosion increases, and hedges undoubtedly provide some protection against that. Much has been claimed about the value of hedgerows in this respect. The truth of the matter is that they afford protection on their lee side for a distance of only between 3·5 and 6 metres from the hedge. This means that for a hedge of average height complete protection would be afforded to fields of no more than about a tenth of a hectare or a third of an acre. The value of the ordinary farm hedge as a means of erosion control around large arable fields is therefore very slight. Nevertheless, there are situations where hedges are

grown specifically as high shelter belts around particular crops like the apple orchards and hop gardens of Kent and the Vale of Evesham, and their importance remains as great as ever.

The final episode of hundreds of years of careful management

The actual amount of hedgerow lost as a direct consequence of this increase in agricultural efficiency has been a matter of some dispute, but it is now more or less agreed that in England and Wales, for the period 1946–74, the best estimate is about 4,500 miles of hedgerow each year. The rate of loss was not uniform over the whole period, nor was it the same in different parts of the country. As might be expected, the greatest effect was in parts of eastern England, whereas in the South-west there was very little change. For instance, about half of Norfolk's hedges may have gone but only a tenth of those in Devon. As a square mile of arable land in East Anglia may originally have contained about ten to fifteen miles of hedge and the South-west, twenty-five to thirty miles, the effect in Norfolk has been dramatic but scarcely noticeable in Devon. Pollard, Hooper and Moore have calculated in more detail the rate of hedge loss for the six million acres of East Anglia, which includes most of the major cereal-producing area of the country. Removal was fairly slow immediately after the war, averaging about 800 miles per year from 1946–54; the rate then accelerated to 2,400 miles per year up to 1962 and reached a peak during the following four years of about 3,500 miles per year. After that the rate of loss declined to about 2,000 miles per year up to 1970.

146

It must be remembered that farmers are not alone responsible for the decline in our hedgerows. Land taken out of agriculture for development and motorway construction also accounts for some losses, albeit relatively small in comparison with those attributable to agricultural changes.

Between 1957 and 1970, some of the hedgerow removal was carried out with the assistance of Ministry of Agriculture grants under the Farm Improvement Scheme, but the actual amount of money paid out was probably quite small. For one thing, only about a fifth of the hedge removal received grant aid anyway, and then the grant itself covered only twenty-five per cent of the approved expenditure. Grant aid specifically for hedgerow removal was withdrawn at the end of 1970 when the new Farm Capital Grant Scheme came into operation.

The destruction of hedgerows still continues, although the rate of loss has undoubtedly declined from its peak in the 1960s. In all parts of the country, as we have seen, hedgerows are a particularly rich habitat for wildlife, and for those areas where agriculture is most intensive they represent almost the only cover left – linear nature reserves within a wild-life desert. Although no species of plant or animal is confined to hedgerows so far as we know, twenty-three species of birds – a quarter of the total British breeding species – regularly nest in hedgerows, and almost seventy have bred in hedges at some time or other. Similarly, about a third of all British flowering plants have been recorded from hedgerows, and something like 250 species occur regularly. As we saw last month, hedges are equally important as cover for mammals, and the insect fauna is one of the richest of any habitat in the country.

Altogether, there are about 400,000 miles of hedge in England and Wales. If we assume an average width of hedge plus ditch and bank of six feet, this is equivalent to an area of 327,273 acres or 132,446 hectares. The 113 National Nature Reserves in England and Wales total 38,398 hectares. In other words, in those two countries the area of hedges is three and a half times greater than the area of National Nature Reserves. Happily, the significance of hedgerows for wild-life conservation is now much more widely appreciated. This stems in large part from the public concern that was expressed when the true extent of hedgerow losses began to be appreciated in the late 1960s. However, there is virtually no legislation by which statutory protection can be conferred on individual important hedges that are under threat from whatever cause.

The Nature Conservancy Council (NCC) is obliged to notify the appropriate local authority of Sites of Special Scientific Interest (SSSIs), but all this does is to ensure that the NCC is consulted if any development is proposed affecting the site. Under the new Wildlife and Countryside Act, 1981, farmers are now required to consult the NCC before embarking on any changes in land use that would affect the value of an SSSI. Although many SSSIs include hedgerows, only one hedgerow has so far been scheduled as an SSSI in its own right, and that is an ancient hedge in Dyfed with twenty-one shrub species in its 120-metre length. Cambridgeshire County

The English countryside in summer

Council has recently established that an overgrown hedge that has become a line of trees and a significant feature in the landscape can be subject to a Tree Preservation Order (TPO). However, it is doubtful if a TPO could be applied purely on biological or historical grounds.

A seminal conference held at Silsoe in Bedfordshire in 1969 brought together farmers and conservationists for the first time to discuss the conflicts of interest between modern agriculture and conservation, and out of this arose the Farming and Wildlife Advisory Group (FWAG), jointly sponsored by the voluntary conservation organisations and the farming industry. Many counties now have their own local FWAG organisations which advise

148

farmers on ways of reconciling agricultural and conservation objectives on their land.

The Countryside Commission's Demonstration Farm Project is breaking new ground in demonstrating how modern agriculture can be pursued with the minimum impact on traditional landscapes, and many local authorities have followed suit with their own projects with the help of Commission finance.

Straightforward grubbing out is not the only way that hedges can be damaged. The stubble burning that occurs after harvest has set fire to many a hedge, but the danger from this is minimal if the National Farmers' Union's own code of practice on straw and stubble burning is conscientiously followed.

It is easily forgotten that there are today many instances where new hedgerows are being planted. We met a fruit farmer who was planting shelter hedges of grey alder around his Gloucestershire vineyards. Highway authorities are planting hedges as accommodation works along new carriageways and road improvements, and some of the farms in the various demonstration projects are establishing new hedgerows.

Under the present Farm Capital Grant Schemes financial assistance is available not only for planting new hedges but also for the long-term renovation of neglected ones. The Government's Agricultural Development and Advisory Service has recently published two new leaflets, 'Planting Farm Hedges' and 'Managing Farm Hedges', and it is to be hoped that all this will encourage farmers to look with more favour on their hedges when their usefulness comes under scrutiny.

The hedgerow landscape is one of the most typical, beautiful and evocative of all English scenes, and something that most people wish to see cherished. Many hedges are genuine archaeological monuments, and some hedge landscapes are almost pure thirteenth century or earlier. Furthermore, landscapes such as this are inevitably getting scarcer. Scenes that are at the same time beautiful, emotive, ancient and rapidly disappearing are surely worthy of a nation's concern.

The trouble is that many of these attributes are intangibles, impossible to price. However, if society decides that our traditional landscapes should be cherished, it is right that the individual landowners, whose responsibilities they ultimately are, should not be financially disadvantaged in the common cause. This is the dilemma. But it has to be resolved, or we shall live to witness John Stuart Mill's prophetic vision: 'Nor is there much satisfaction in contemplating a world with every rood of land brought into cultivation, every flowery pasture ploughed up, every hedgerow or superfluous tree rooted out and scarcely a place left where a wild shrub or flower could grow.'

Winter twigs

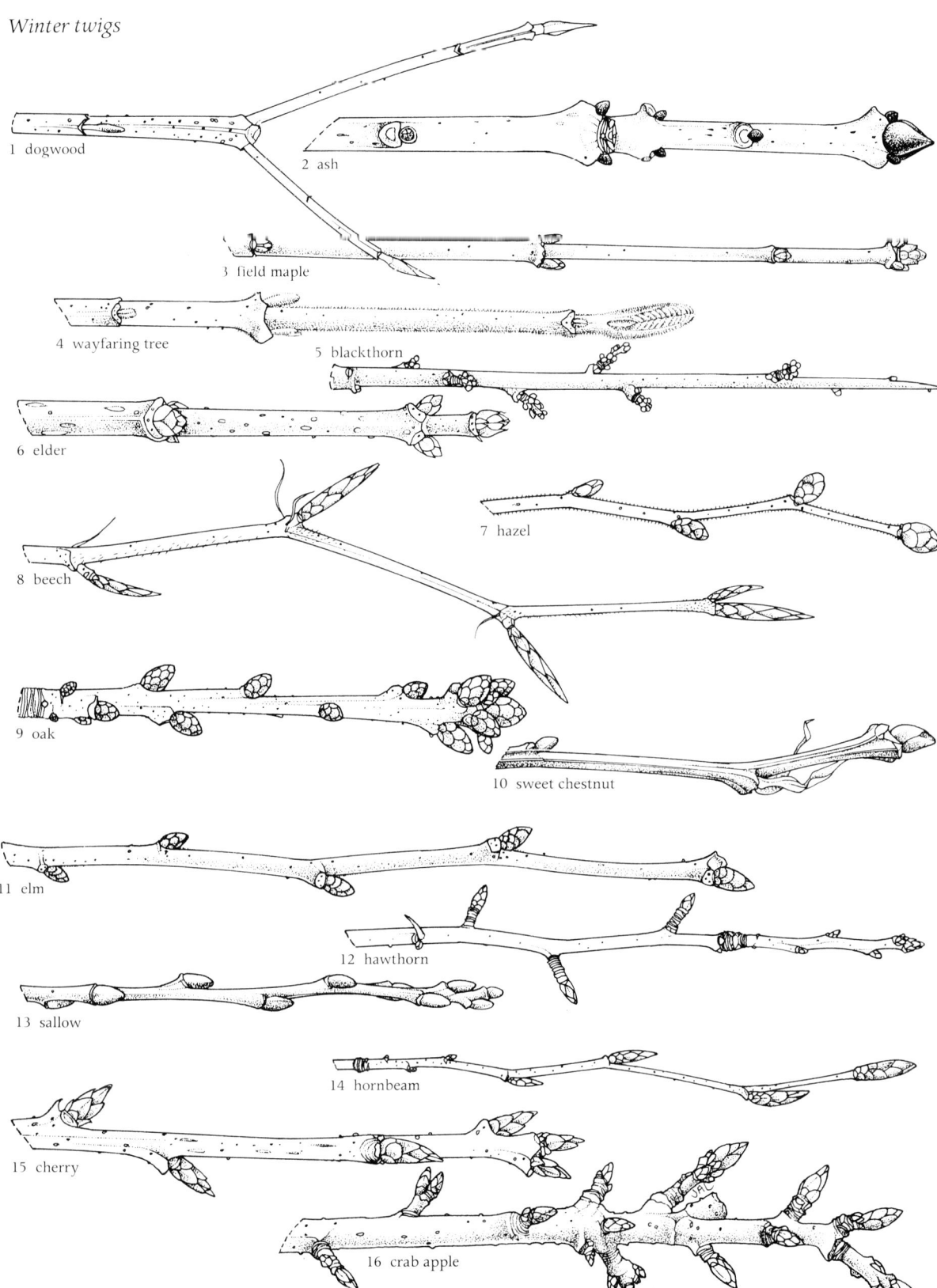

1 dogwood

2 ash

3 field maple

4 wayfaring tree

5 blackthorn

6 elder

7 hazel

8 beech

9 oak

10 sweet chestnut

11 elm

12 hawthorn

13 sallow

14 hornbeam

15 cherry

16 crab apple

Key to the twigs of the commoner deciduous hedgerow trees and shrubs

1	Buds in opposite pairs on either side of the stem		2
	Buds not opposite		9
2	Bark of young twigs dark shining red	Dogwood	
	Bark of young twigs not a dark shining red		3
3	Bark of young twigs bright green	Spindle	
	Bark of young twigs not bright green		4
4	Buds black, leaf scars large, twigs stout and grey	Ash	
	Buds not black		5
5	Buds downy, without bud-scales	Wayfaring tree	
	Buds with bud-scales		6
6	Bark pale and warty, twigs with large white pith	Elder	
	Twigs without large white pith		7
7	Twigs thorny, wood orange beneath bark	Buckthorn	
	Twigs not thorny		8
8	Buds green, leaf scars large	Sycamore	
	Buds not green, leaf scars very small	Field maple	
9	Twigs thorny and/or spiny		10
	Twigs neither thorny nor spiny		11
10	Young twigs reddish-brown with numerous sharp lateral spines	Hawthorn	
	Young twigs grey-black, dull and downy, tips of short lateral branches spine-like	Blackthorn	
11	Young twigs downy (*examine carefully*)		12
	Young twigs not downy (*examine carefully*)		13

12	Buds greenish, more than 3 mm long	Hazel	
	Buds dark brown, less than 3 mm long	Elm	
13	Twigs with short wrinkled 'spur' shoots		**14**
	Twigs without short wrinkled 'spur' shoots		**15**
14	Buds with hairy bud-scales; bark not peeling horizontally	Crab-apple	
	Buds without hairy bud-scales; bark smooth and shining; peeling horizontally. Twigs with lenticels (small pale warty spots)	Wild cherry	
15	Buds spindle-shaped, more than three times as long as broad		**16**
	Buds less than three times as long as broad		**17**
16	Buds more than four times as long as broad	Beech	
	Buds less than four times as long as broad	Hornbeam	
17	Young twigs and buds glossy reddish brown; twigs with prominent longitudinal ridges	Sweet chestnut	
	Twigs without prominent longitudinal ridges		**18**
18	Buds with a single bud-scale, flattened, lying approximately flush with twig	Sallow	
	Buds not as above		**19**
19	Buds dark red and glossy	Lime	
	Buds not dark red		**20**
20	Buds purplish and dull	Alder	
	Buds not purplish and dull		**21**
21	Young twigs very slender, dark brown, glossy with small warts	Silver birch	
	Not as above, buds stout, light brown, crowded at tips of twigs	Oak	

Use the key in the same way as that described for the Key to the Commoner Grasses of the Hedgerow (see page 64). In particular, resist the temptation to try and identify the twigs from the illustrations alone and read the hints on page 142.

Bibliography

For anyone seriously interested in hedgerows, the following two books are indispensable reading:

BRITISH TRUST FOR CONSERVATION VOLUNTEERS *Hedging: a practical conservation handbook* Zoological Gardens, London: BTCV, 1975.

POLLARD, E., HOOPER, M.D. and MOORE, N. W. *Hedges* (New Naturalist) Collins, 1974

References

BAIRD, W. W. and TARRANT, J. R. *Hedgerow destruction in Norfolk, 1946–70* School of Environmental Studies, University of East Anglia, 1973.

BRANDON, P. F. *Medieval clearances in the East Sussex Weald* in *Institute of British Geographers Transactions* no. 48, December 1969. pp 135–153.

CAMERON, R. A. D., DOWN, K. and PANNETT, D. J. *Historical and environmental influences on hedgerow snail faunas* in *Biological Journal of the Linnean Society* vol. 13, no. 1, 1980. pp 75–87.

HOOPER, M. D. *Dating hedges* in *Area* no. 4, 1970. pp 63–65.

HOOPER, M. D. *Hedgerow removal* in *Biologist* vol. 21, no. 2, 1974. pp 81–86.

MOORE, N. W., HOOPER, M. D. and DAVIS, B. N. K. *Hedges I. Introduction and reconnaissance studies* in *Journal of Applied Ecology* vol. 4, no. 1, May 1967. pp 201–220.

NATURE CONSERVANCY *Hedges and hedgerow trees* Monks Wood Symposium 4, 1968.

POLLARD, E. *Hedges VII. Woodland relic hedges in Huntingdon and Peterborough* in *The Journal of Ecology* vol. 61, 1973. pp 343–352.

POLLARD, E. and RELTON, J. *Hedges V. A study of small mammals in hedges and cultivated fields* in *Journal of Applied Ecology* vol. 7, no. 3, December 1970. pp 549–557.

PULLEIN, C. *Rotherfield: the story of some Wealden manors* Tunbridge Wells: Courier Printing and Publishing Co., 1928.

SOUTHWOOD, T. R. E. *The number of species of insect associated with various trees* in *Journal of Animal Ecology* vol. 30, 1961. pp 1–8.

Free information sheets

MINISTRY OF AGRICULTURE Agricultural Development and
Advisory Service *Managing farm hedges* and *Planting farm hedges*
(Leaflets 762 and 763) ADAS, 1980.

Further reading

BAKER, M. *Discovering the folklore of plants* Shire Publications,
n.e. 1980.

BASELEY, G. *A country compendium* Sidgwick and Jackson, 1977;
Star Books, n.e. 1980.

BRANDON, P. *Sussex* (The making of the English landscape)
Hodder, 1974.

GRIGSON, G. *The Englishman's flora* Hart-Davis, 1975; Paladin,
1975.

HATFIELD, A. W. *How to enjoy your weeds* Muller, n.e. 1974.

HOSKINS, W. G. *English landscapes* BBC Publications, n.e. 1977.

HOSKINS, W. G. *The making of the English landscape* Hodder, n.e.
1977; Penguin Books, 1970.

HYDE, M. *Hedgerow plants* Shire Publications, 1976.

LOEWENFELD, C. and BACK, P. *Britain's wild larder* David and
Charles, 1980.

MICHAEL, P. *All good things about us* Benn, 1980.

RACKHAM, O. *Trees and woodlands in the British landscape* Dent,
n.e. 1981.

RICHARDSON, R. *Hedgerow cookery* Penguin Books, 1980.

YELLING, J. A. *Common field and enclosure in England 1450–1850*
Macmillan, cased and paperback 1977.

Plant names

Throughout the book English names of plants are those
recommended in *English names of wild flowers* edited by J. G.
Dony et al, and published by Butterworth for the Botanical
Society of the British Isles, 1974. Scientific names follow those of
Flora Europaea edited by T. G. Tutin et al. 4 vols. Cambridge U.P.,
1964–76.

Acknowledgements

AEROFILMS LTD Laxton fields page 12; AGRICULTURAL RESEARCH COUNCIL, WEED RESEARCH ORGANISATION, KIDLINGTON rye grass, crested dog's tail & cock's foot grass page 60, nettle page 77; HEATHER ANGEL peacock pupa & butterfly page 90, oak galls page 97, wood mouse page 126, hedgehog page 132, sloes & haws page 143, hedgerows page 144; AQUILA PHOTO-GRAPHICS magpie page 41, bee page 76, maple galls page 97, white briony page 100; ARDEA hedge sparrow page 36, wild strawberries (photo Ake Lindau) page 72, grubbing up hedge (photo E. Mickleburgh) page 146; BARNABY'S PICTURE LIBRARY blackbird (Mustograph) page 40, hedgerow tree (Mustograph) page 119, cut & layered hedge (photo John H. Gardner) page 138, trimming hedge (photo Tony Boxhall) page 139; J. ALLAN CASH hedge page 9, Devon landscape page 148; BRUCE COLEMAN LTD yellow hammer (photo R.K. Murton) page 36, whitethroat (photo Dennis Green) page 36, comfrey (photo Hans Reinhard) page 54, brimstone butterfly (photo Jane Burton) page 90, dormouse (photo K. Weber) page 126, badger (photo Hans Reinhard) page 132, spindle fruit (photo Peter Ward) page 143, hips (photo Jane Burton) page 143, pleached hedge (photo Mark Boulton) page 144, fieldfare (photo G. Langsbury) page 144; EAST SUSSEX COUNTY RECORD OFFICE map page 58; ESSEX COUNTY RECORD OFFICE sur-veyors & two maps all page 57; DAVID HOSKING robin page 18, chaffinch page 36, bank vole & common shrew page 126; LINNET PHOTOGRAPHIC (photo Ivor J. Dixon) floods page 120; JAMES RAVILIOUS/THE BEAFORD ARCHIVE laying hedge page 8, hedging page 138; DAVID STREETER Sussex hedge, catkins & Essex hedge all page 18, butterfly page 36, cuckoo spit, tufted vetch, dog rose & common vetch all page 54, caterpillars, soldier beetle & horsefly all page 72, comma butterfly, speckled wood butterfly & hedge brown butterfly all page 90, all five pictures page 108, salmon traps page 120; JOHN TOPHAM dry stone walls page 9, violets page 28, wood anemone page 31, wren page 40, thrush page 41, comfrey page 62, weasel page 130, stoat page 132; UNIVERSITY OF CAMBRIDGE fields page 14; THE WHITBY GAZETTE 'Penny hedge' page 10.

Acknowledgment is also due to:
WILLIAM COLLINS SONS & CO LTD for illustration from *Hedges* by Pollard, Hooper and Moore.

Index